Synthesis and Properties of POLYMERS

Synthesis and Properties of POLYMERS

G.E. ZAIKOV (Editor)

Nova Science Publishers, Inc.
Commack

Art Director: Maria Ester Hawrys
Assistan Director: Elenor Kalberg
Graphics: Denise Dieterich , Kerri Pfister,
Barbara Cassatti and Erika Minerd
Manuscript Coordinator: Sharyn Schweidel
Book Production: Tammy Sauter and Benjamin Fung
Circulation: Irene Kwartiroff and Annette Hellinger

Library of Congress Cataloging-in-Publication Data

Synthesis and properties of polymers/Zaikov, G.E. (ed.).
p. cm.
Includes bibliographical references and index.
ISBN 1-56072-257-6:
1. Polymers. 2. Polymerization. I. Zaikov, Gennadii Efremovich.
QD381.7 S95 1995 95-24678
547.7'.0459--dc20 CIP

6080 Jericho Turnpike, Suite 207
Commack, New York 11725
Tele. 516-499-3103 Fax 516-499-3146
E Mail Novasci1@aol.com

Printed in the United States of America

CONTENTS

A New Approach to the Description of the Mechanical Behavior of Polymer Glasses

M.S. Arzhakov and S.A. Arzhakov
Polymer Department, Faculty of Chemistry,
Lomonosov Moscow State University,
Vorob'evy Gory, Moscow, 119899 Russia

Temperature-induced relaxation of plastic deformation of polymer glasses based on PMMA was studied. Plastic deformation of polymers and their mechanical properties were described in terms of contributions from low-temperature and high-temperature components to the recovery of residual deformation. The above components of relaxation were treated as physical parameters of deformation and relaxation in polymer glasses. The ratio between low-temperature and high-temperature components was shown to be controlled by physicochemical modification of polymer materials.

Introduction

Temperature-induced relaxation of plastic deformation of polymer glasses is known to take place in two well-defined temperature regions [1-5]. A complete recovery of geometrical sizes of plastically deformed polymers is observed in glass transition region. However, a marked fraction of plastic deformation is shown to relax at temperatures much below the glass transition point. In other words, relaxation of plastic deformation involves two modes: low temperature and high-temperature modes. The studies on mechanical behavior of glassy polymers under various modes of deformation (uniaxial compression and drawing, hydrostatic compression, and creep) show [3-5] that the two-stage character of relaxation of plastic deformation is a universal feature of all polymer bodies.

The complicated character for relaxation of plastic deformation of glassy polymers is a manifestation of physical processes, which are responsible for and accompany plastic deformation. In this case, the mechanical properties of polymer bodies seems to be controlled by coexistence of the above modes of relaxation.

The aim of this work is to describe the mechanical behavior of polymer glasses in terms of low-temperature and high-temperature components for relaxation of plastic deformation in glassy polymers.

EXPERIMENTAL

Commercial PMMA samples were used. Dibutyl phthalate (DBPh) was used as the plasticizing agent. The content of plasticizing agent was varied from 1 to 10 mass%. The copolymers of MMA with bifunctional monomer ethylene glycol dimethacrylate (EGDM) were used as cross linked PMMA. The development of network took place immediately during bulk copolymerization. The content of EGDM in the polymerization system was 1.0, 5.2 and 11.2 mol%.

The test samples were cylinders with a height of 30 mm and a diameter of 15 mm. Prior to testing, polymer samples were annealed at temperatures above T_g by 10-15° and then they were slowly cooled down to room temperature.

The samples of initial and plasticized PMMA were uniaxially compressed at temperatures 293, 320, 333, 343, 353, and 383 K with a strain rate of 0.1 mm/mn using an UTS-100 tensile machine (Germany). After that, they were cooled with liquid nitrogen in a stressed state. Then, the free-standing samples were heated from 213 K at a heating rate of 1 K/min. The relaxation (recovery) of residual plastic deformation was estimated by measuring the height of the samples with an accuracy of ±0.001 mm.

The samples of the cross-linked PMMA were uniaxially compressed at 293 K and then they were unloaded with the same rate at this temperature. The relaxation of residual deformation was studied starting with 293 K.

Elastic modulus was estimated from the initial slope of the σ-ε curve at strains below 3%.

RESULTS AND DISCUSSION

A typical temperature dependence of recovery (relaxation) of residual deformation ε_{res} in PMMA samples is presented in Figure 1. This curve shows that ε_{res} involves low-temperature (ε_1) and high-temperature (ε_2) components and may be represented as follows

$$\varepsilon_{res}=\varepsilon_1 + \varepsilon_2 \quad (1)$$

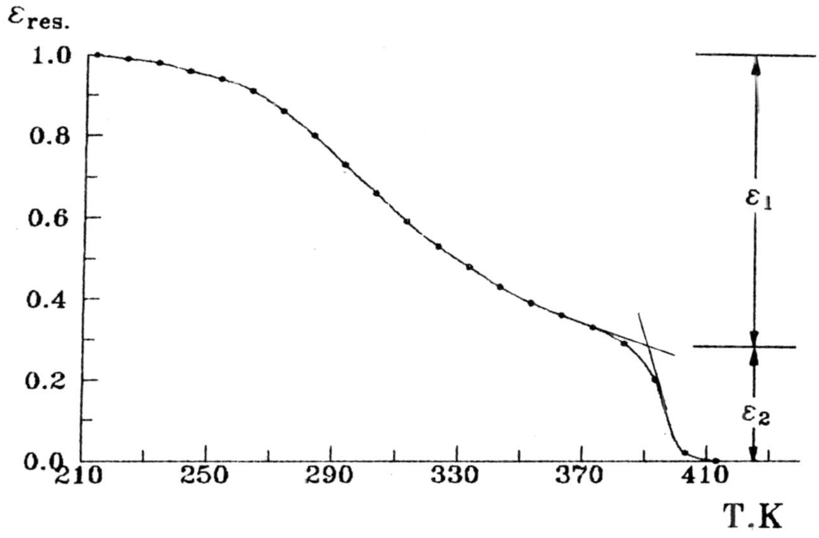

Figure 1. Typical curve of recovery of residual deformation ε_{res} in PMMA sample. Strain: 20%. T_{def}=293K.

Note that relaxation of ε_{res} via high-temperature mode ε_2 is observed in rather narrow temperature region of glass transition. Relaxation of plastic deformation via low-temperature component ε_1 takes place in a wide temperature region below glass transition point, and each temperature is associated with a certain portion of recovery of ε_{res}.

The ratio between the components associated with low-temperature and high-temperature relaxation was studied as a function of strain. For the compressed samples with rather low strains (Figure 2, curves 1, 2), complete strain recovery takes place at temperatures below T_g. On the contrary, a complete strain recovery of the compressed samples with high strains is observed on heating to T_g (Figure 2, curves 3-5). Note that in this case low-temperature relaxation is also involved.

Figure 3 presents low-temperature (ε_1) and high-temperature (ε_2) components of relaxation of plastic deformation as a function of strain and the corresponding stress-strain curve of uniaxial compression of PMMA samples. As follows from Figure 3, in the case of the samples with strains below yield strain ε_y, relaxation of plastic deformation proceeds via low-temperature mode ε_1. At the strain corresponding to ε_y, component ε_1 is maximal, and further increase in compression strain is associated with the appearance and growth in component ε_2 at constant ε_1.

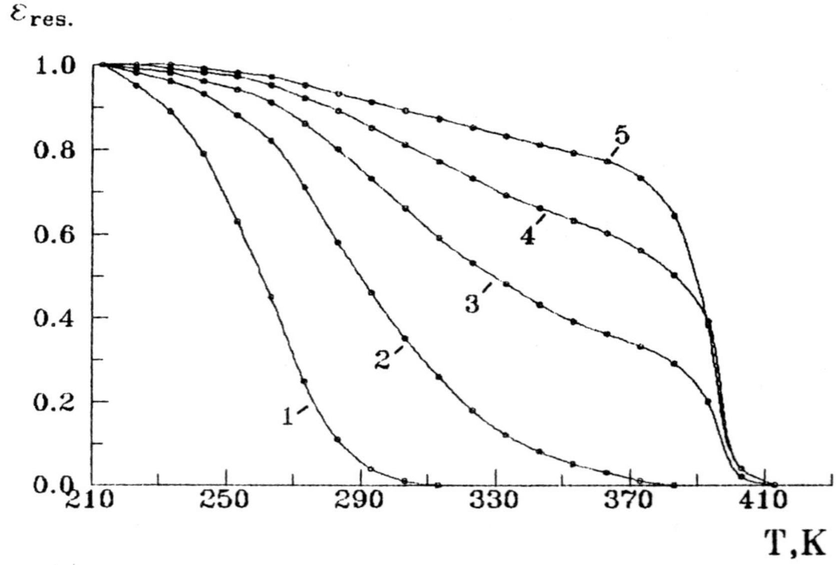

Figure 2. Recovery of residual deformation ε_{res} in PMMA samples with strains 10 (1), 14 (2), 20 (3), 30 (4), and 50% (5). T_{def}=293K.

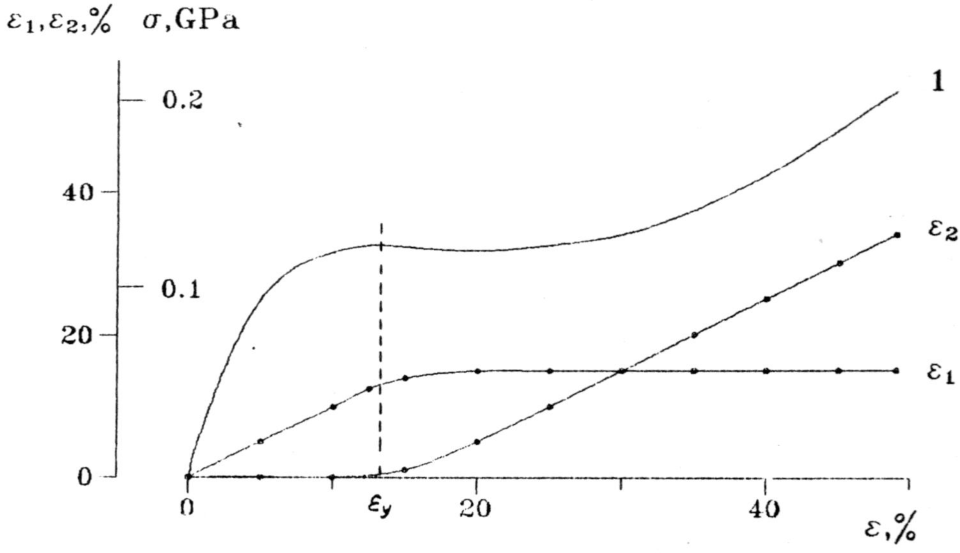

Figure 3. Stress-strain curve corresponding to uniaxial compression of PMMA samples at 293 K (curve 1) and low-temperature ε_1 and high-temperature components ε_2 of strain recovery versus strain ε.

At the present time, the two-stage character of recovery of polymer plastic deformation is treated in terms of two models: step-by-step recovery of small-scale plastic shear transformations and excited chain conformations [4,5] (i) and structural heterogeneity of glassy polymers [3] (ii).

Within the framework of the latter approach, glassy polymer is characterized by a set of structural sublevels with different packing densities and ordering. Let us note that structural heterogeneity in glassy polymers is widely discussed in literature [6-9].

In our opinion, each of these sublevels is associated with its own relaxation time and activation energy of segmental motion. Hence, at each temperature below glass transition point local segmental mobility in certain structural sublevels is allowed. As temperature increases, segmental mobility in structural levels with higher packing density appears, and finally at T_g all polymer sample is involved in segmental mobility.

In the case of the deformed polymer samples, as temperature increases, the appearance of segmental mobility in structural levels with higher packing density is accompanied by a gradual recovery of local plastic deformation within the above structural sublevels. (In addition, the appearance of local segmental mobility at temperatures below T_g is likely to be controlled by stored inner stresses. As a result, effective energy of activation of segmental motion decreases, and segmental relaxation at local structural sublevels is allowed.) Gradual appearance of segmental mobility in structural levels with higher packing density allows one to explain monotonous temperature dependence of recovery of ε_{res} (Figure 1, component ε_1) as well as the fact that each temperature is associated with a certain value of strain recovery. A complete recovery of plastic deformation is accomplished at T_g, when all segments are involved in segmental motion (Figure 1, component ε_2). Hence, molecular mechanism of low-temperature relaxation is similar to that of high-temperature mode and involves transition of excited chain conformations to initial equilibrium state via segmental motion.

So, in the case of glassy polymers with low strains below or equal to ε_y, residual deformation is controlled by low-temperature relaxation via local segmental mobility at structural levels with lower packing density (component ε_1). This implies that polymer deformation to low strains below ε_y is associated with local segmental mobility within certain structural levels.

As strain increases above ε_y, component ε_2 increases. This mode of relaxation is observed in glass transition region and realized via segmental mobility of polymer sample in whole. This suggests that at such strains segments of all polymer sample are involved in plastic deformation.

Relaxation of plastic deformation at temperatures below T_g (component ε_1) allows one to conclude that relaxation of this portion of plastic deformation is allowed even during polymer loading. Let us define this component as ε_1^*. From this viewpoint, component ε_1 stands for stored plastic deformation, which did not relax during polymer loading. Furthermore, a certain fraction of component ε_1 is lost on heating polymer sample unloaded in liquid nitrogen to initial test temperature (213 K). Nevertheless, with an accuracy to this value plastic deformation of glassy polymer may be presented as

$$\varepsilon = e_1^* + \varepsilon_{res} = (\varepsilon_1^* + \varepsilon_1) + \varepsilon_2 \qquad (2)$$

(In this case, Hookean components of deformation and relaxation are ignored.)

According to equation (2), plastic deformation of glassy polymer may be described by the contributions from ε_1 and ε_2 to relaxation of residual deformation. At $\varepsilon \leq \varepsilon_y$, deformation is primarily controlled by $(\varepsilon^*{}_1 + \varepsilon_1)$, whereas at $\varepsilon \geq \varepsilon_y$ component ε_2 comes into play.

With respect to polymer strength properties, consideration of component $(\varepsilon_1^* + \varepsilon_1)$ is of most interest. This parameter is responsible for the development of polymer deformation at initial portion of σ-ε curve, i.e., at $\varepsilon \leq \varepsilon_y$, which is associated with such mechanical properties as elastic modulus E_o and yield stress σ_y. This allows one to advance the correlation between relaxation $(\varepsilon_1^* + \varepsilon_1)$ and mechanical characteristics E_o and σ_y.

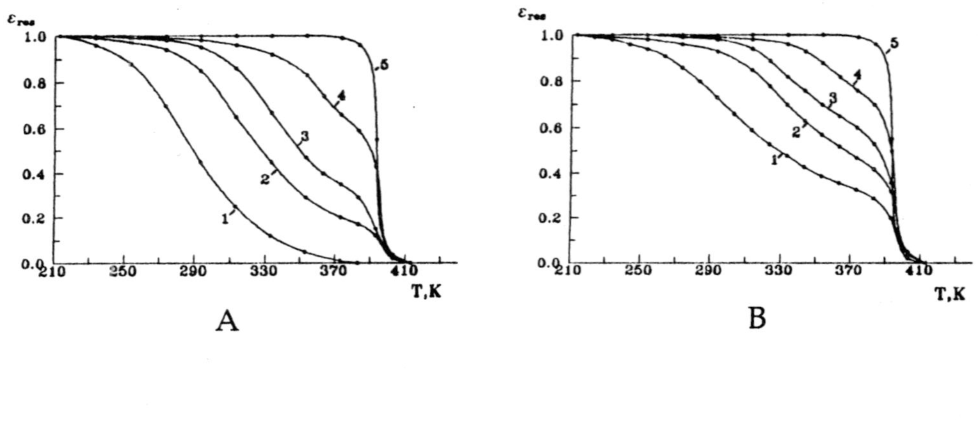

Figure 4. Recovery of residual deformation eres in PMMA samples at T_{def}=293 (1), 320 (2), 333(3), 353 (4), and 383K (5). Strain 14(A) and 20% (B).

To verify this correlation, PMMA samples were compressed to different strains at different temperatures. Then, relaxation of ε_{res} was studied (Figure 4). For the lack of space, recovery curves are presented only for the samples with 14 and 20% strains. As T_{def} increases, the contribution from component ε_1 to strain recovery is seen to decrease. At test temperatures close to glass transition point, the value of ε_{res} is controlled only by component ε_2 (Figure 4, curves 5). Furthermore, as T_{def} increases, elastic modulus and yield point go down.

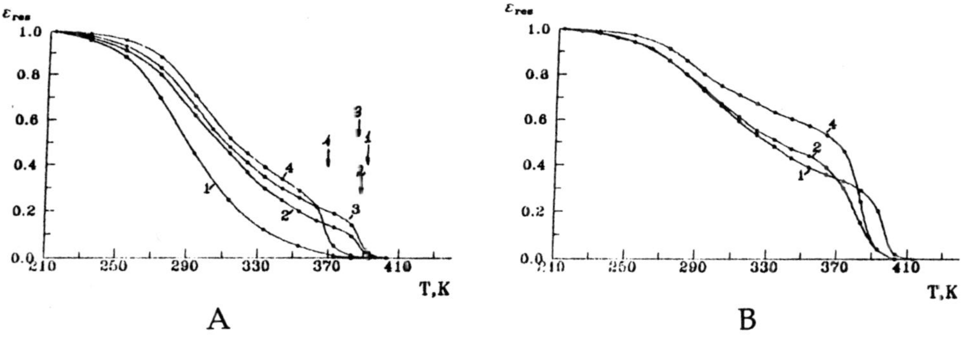

Figure 5. Recovery of residual deformation ε_{res} of plasticized PMMA samples. Content of plasticizing agent DBPH: 0(1), 1(2), 5 (3), and 10 mass% (4). Strain: 14 (A) ad 20% (B). T_{def}=293K. The arrows show T_g of the corresponding polymers.

The similar results were observed for deformation and relaxation of PMMA samples plasticized with DBPh (Figure 5). The effect of plasticizing agent on the ratio between ε_1 and ε_2 is similar to increasing deformation temperature T_{def}. Moreover, experiments revealed that at deformation temperatures equidistant from glass transition point both initial and plasticized samples showed the same ratio between ε_1 and ε_2.

This experimental evidence suggests an intimate correlation of elastic modulus and yield stress with low-temperature component of relaxation of plastic deformation (Figure 6). At any strains, similar tendency is observed: as contribution from component ε_1 to recovery of ε_{res} decreases, both elastic modulus and yield stress tend to decrease. In this case, the way of varying the above parameters [by increasing T_{def} (circles in Figure 6) or content of plasticizing agent (squares in Figure 6B)] has no importance.

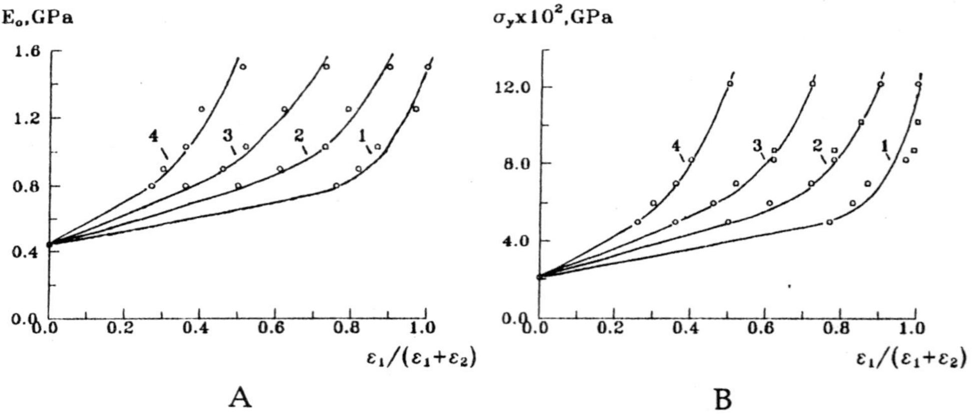

Figure 6. The dependence of elastic modulus E_o (A) and yield stress σ_y (B) of PMMA samples on the contribution of low-temperature component ε_1 to total recovery of residual deformation ε_{res}. Strain: 10 (1), 14 (2), 20 (3), and 30% (4).

Let us note that the dependences of E_o and σ_y on contribution from component ε_1 to recovery of ε_{res} show a well-pronounced inflection at temperatures from 343 to 353K. The dependence of abscissa of this inflection point $[\varepsilon_1/(\varepsilon_1+\varepsilon_2)]^*$ on strain ε is linear (Figure 7). The origin of such behavior is beyond the scope of this work and will be discussed in the following paper.

A noticeable decrease in E_o and σ_y with decreasing component ε_1 is rationalized by the fact that as T_{def} or content of plasticizing agent in polymer increases, the fraction of plastic that relaxes during polymer deformation increases. In terms of equation (2), this implies an increase in component ε_1^* via decreasing component ε_1. In turn, as contribution from component ε_1 to recovery of ε_{res} decreases, component ε_2 increases. The increase in component ε_1^* associated with relaxation during loading is accompanied by a concomitant decrease in E_o and σ_y of glassy polymer, for these parameters are controlled by molecular dynamics of polymer macrochains.

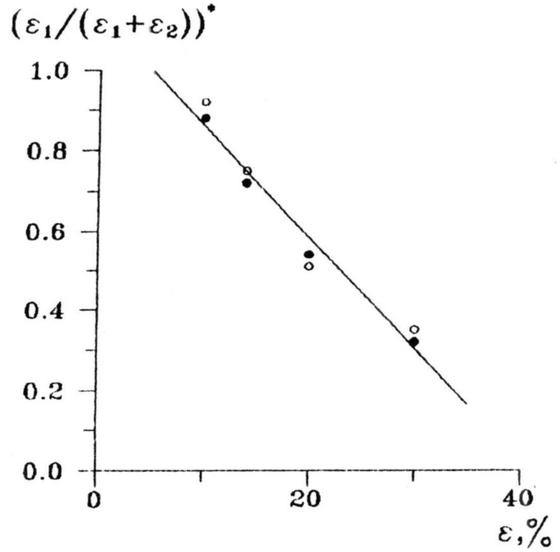

Figure 7. The dependence of abscissa of inflection point $[\varepsilon_1/(\varepsilon_1+\varepsilon_2)]^*$ in curves $E_o=f[\varepsilon_1/(\varepsilon_1+\varepsilon_2)]$ and $\sigma_y=f[\varepsilon_1/(\varepsilon_1+\varepsilon_2)]$ on strain ε as obtained from curves $E_o=f[\varepsilon_1/(\varepsilon_1+\varepsilon_2)]$ (closed circles) and $\sigma_y=f[\varepsilon_1/(\varepsilon_1+\varepsilon_2)]$ (open circles).

As T_{def} approaches T_g, ε_1 tends to zero, and residual deformation is controlled only by component ε_2 (Figure 4, curves 5). In this case, elastic modulus and yield stress of PMMA samples approach the values, which are characteristic of rubbery state.

Hence, the ratio between components ε_1 and ε_2 to recovery of plastic deformation provides an important information concerning specific features of polymer deformation and relaxation and is responsible for polymer mechanical behavior. The above values may be treated as new physical parameters of deformation of polymer glasses.

Correlation between the above parameters and polymer brittleness is quite evident. As was shown earlier, as T_{def} decreases, component ε_2 tends to vanish (Figure 4). Brittle fracture of polymer material is likely to take place at temperatures, where component ε_1 is zero. Degeneration of component ε_1 implies that segmental mobility in polymer samples is ceased at all levels of structural inhomogeneity. Hence, temperature region from T_{br} to T_g may be treated as temperature interval, where higher limit is associated with manifestation of segmental mobility in whole polymer sample, and lower limit corresponds to frozen-in segmental mobility in whole polymer sample.

Furthermore, studying the contributions of components ε_1 and ε_2 to recovery of plastic deformation of polymer glasses is of primary importance from the viewpoint of polymer processing into final articles. Component ε_1 is the main reason of such common phenomenon as low shape stability of deformed or oriented polymer samples at temperatures much below polymer glass point. In turn, this low-temperature shrinkage is associated with dramatic deterioration in mechanical properties of polymer articles. In connection with this, preventing this mode of polymer relaxation is of primary importance.

The data presented in Figure 4 suggests that preventing relaxation component ε_1 is provided by increasing deformation temperature T_{def}. This approach is used as a basis for traditional technology of preparation of orientation-strengthened polymer glasses. This technology involves polymer orientation (tensile drawing) at temperatures above glass transition point and further cooling in a stressed state. Materials and articles prepared via this method are characterized by rather high shape and orientation stability and, hence, by high mechanical properties at temperatures below T_g.

On the other hand, component ε_1 is responsible for low-temperature plasticity of polymer bodies and high reversible strains, which are known to be valuable properties of polymer plastics. This important feature of deformation and relaxation behavior of polymer glasses seems to be intimately related to their unique properties, in particular, lowered brittleness as compared with low-molecular-mass glasses.

Hence, mechanical behavior of polymer glasses is mainly controlled by low-temperature component of relaxation of polymer glasses, that is, by the contribution from component ε_1 to recovery of ε_{res}. In connection with this, the problems concerning modification of polymer materials should be treated, in general, as controlled variation in the contribution of component ε_1 in polymer relaxation. In our opinion this approach offers many advantages in controlled modification of polymer materials.

As was shown earlier, increasing T_{def} as well as polymer plasticization is accompanied by a decrease in contribution from component ε_1 to recovery of plastic deformation of PMMA samples. In contrast, cross-linking of PMMA samples with EGDM is associated with increasing contribution from component ε_1 (Figure 8), and at relatively high content of EGDM (5-11 mol%), relaxation of plastic deformation is completed at temperatures below T_g only via component ε_1.

So, the above experimental evidence suggests possibilities, which allow one to control variations in the contribution of ε_1 in relaxation of plastic deformation of polymer glasses via physicochemical modification.

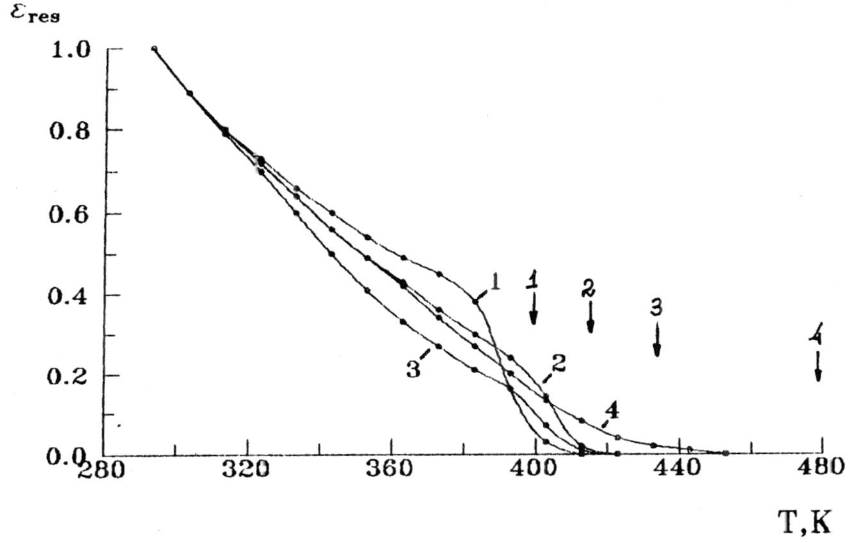

Figure 8. Recovery of residual deformation eres of cross-linked PMMA samples. Content of cross-linking agent (EGDM): 0(1), 1 (2), 5.2 (3), and 11.2 mol.% (4). Strain: 20%. T_{def}=293K. The arrows show T_g of the corresponding polymers.

CONCLUSION

Plastic deformation of glassy polymers may be described by the contributions from low-temperature and high-temperature components temperature-induced relaxation of residual deformation.

Below yield-point, polymer deformation is primarily associated with low-temperature component ε_1. In this case, elastic modulus and yield stress are controlled by contribution from component ε_1 to relaxation of plastic deformation.

In more general sense, the ratio between low-temperature and high-temperature components is responsible for mechanical behavior of polymer glasses: brittleness, elasticity, and plasticity as well as shape stability and shrinkage of oriented polymer glasses. In connection with this, the above ratio between low-temperature and high-temperature components may be treated as a new physical parameter of deformation and relaxation of polymer glasses.

This ratio may be varied by physicochemical modification of polymer materials, i.e., by plasticization and cross-linking. This approach, in our opinion, offers many advantages in controlled modification of the mechanical behavior of polymer materials.

References

1. S.A. Arzhakov and V.A. Kabanov, Vysokomol. Soedin. B13, 318 (1971) (in Russian).
2. A.E. Sorobogatova, S.A. Arzhakov, N.F. Bakeev, and V.A. Kabanov, Dokl. Akad. Nauk SSSR 211, 151 (1973) (in Russian).
3. S.A. Arzhakov, N.F. Bakeev, and V.A. Kabanov, Vysokomol. Soedin. A15, 1154 (1973) (in Russian).
4. E. Oleynik, Progr. Colloid & Polym. Sci. 80, 140 (1989).
5. E.F. Oleinik, O.B. Salamatina, S.N. Rudnev, and S.V. Shenogin, Polym. Sci. 35, 1532 (1993).
6. G.S.J. Yeh, J. Macromol. Sci. B6, 465 (1972).
7. I.J. Klement and P.H. Geil, J. Macromol. Sci. B5, 505 (1971).
8. J.H. Wendorff, Polymer 23, 543 (1982).
9. V.N. Belousov, G.V. Kozlov, A.K. Mikitaev, and Yu.S. Lipatov, Dokl. Akad. Nauk SSSR 313, 630 (1990) (in Russian).

Structural and Mechanical Aspects of Interaction of Polymers with Low-Molecular-Mass Substances

M.S. Arzhakov
Polymer Department, Faculty of Chemistry,
Lomonosov Moscow State University,
Lenin Hills, Moscow, 119899 Russia

Specific features of complexation between low-molecular-mass compounds (LMs) and polymers were discussed in terms of the structural inhomogeneity of amorphous glassy polymers. Introduction of LMs into the polymers was shown to be associated with distribution of the LMs between two different forms, which are characterized by different energetics of bonding with polymer matrix. This factor was responsible for the complicated character of molecular dynamics of the polymer/LM system and mechanical behavior of such complex materials on the whole. Cases when macroscopic mechanical behavior of the polymer containing LM was primarily controlled by the form, in which this compound existed in polymer matrix, were considered. The future development of this direction of polymer modification was discussed and possibilities of its practical application for controlled improvement in the mechanical response of polymer materials were outlined.

Introduction

In many cases physical modification of polymer materials may be achieved by introduction of compatible low-molecular-mass additives (plasticizing agents, dyes, reinforcing agents, flame-retardants, etc.) into polymers. Compatibility of the components is controlled by nonchemical interaction between molecules of low-molecular-mass compounds (LMs) and polymer chains. Such interaction is provided by hydrogen bonding, electrostatic, dispersion, and similar interactions, which, in the general case, are responsible for complexation between the components of polymer composites. Physical and mechanical characteristics of the as-modified polymer materials are decided, within the first approximation, by energies of the interaction between polymer chains and molecules of modifying agent.

In some cases, LM molecules are known to be distributed within the polymer matrix between two different forms, which are characterized by different energy of the interaction with polymers [1-4]. As a result, the physical and mechanical behavior of such complex polymer/LM systems shows a number of peculiar features. In particular, one and the same modifying agent may have quite an opposite effect on mechanical properties of the polymer depending on the form, in which it exists in the polymer matrix [3,5].

In connection with this, studies on the distribution of LM in a polymer matrix between two forms are of great interest and will allow one to gain, on one hand, a deeper insight into the mechanism of this phenomenon and, on the other hand, to take advantage of this phenomenon for controlled modification of the polymer materials.

The aim of the present work is to summarize the published data on the interaction of LMs with polymers. The principle idea of this work is not to cover all the related aspects but to attract the attention of scientists to this fascinating problem and to outline general ways of its further development.

Structural Features of Polymer Interaction with Low-Molecular-Mass Compounds

In some cases, introduction of low-molecular-mass compounds (LMs) into the polymer materials is accompanied by distribution of LM between two different forms, which are characterized by different energy of interaction with polymer matrix. This phenomenon was described in publications devoted to modification of PA-6 with oxyaromatic compounds (OACs) [1,2], sorption of Br_2 by PAN fibers from aqueous bromine solutions [3], transport of dioxydine in ternary system based on dioxydine, water and copolymer of N-vinylpyrrolidone with methyl methacrylate [4].

In all cases, characteristic features of this phenomenon are the following. LMs, which were introduced into the polymer matrix from the swelling solutions (in all cases studied, aqueous solutions were used), appear to be distributed within the polymer matrix between two different forms, which are characterized by reversible or irreversible binding with polymer matrix. Hereinafter, such forms are referred to as reversibly (RB) and irreversibly bound (IB) states. As compared with IB form, RB form may be easily removed from the polymer matrix by keeping the polymer samples in a pure solvent (water).

The common feature of all the above polymer systems is related to the fact that LMs show a high affinity to polymer matrices. Thermodynamical studies of sorption of Br_2 by PAN fibers [3] and OAC by PA-6 films [2] revealed that partition coefficients of LMs between polymer matrices and aqueous media are rather high, that is, preferential sorption of LMs from their aqueous solutions is observed. In the cases studied, the origin of high affinity between polymers and LMs is related to complexation of molecules of

Structural and Mechanical Aspects of Interaction of Polymers with Low-Molecular-Mass Substances

M.S. Arzhakov
Polymer Department, Faculty of Chemistry,
Lomonosov Moscow State University,
Lenin Hills, Moscow, 119899 Russia

Specific features of complexation between low-molecular-mass compounds (LMs) and polymers were discussed in terms of the structural inhomogeneity of amorphous glassy polymers. Introduction of LMs into the polymers was shown to be associated with distribution of the LMs between two different forms, which are characterized by different energetics of bonding with polymer matrix. This factor was responsible for the complicated character of molecular dynamics of the polymer/LM system and mechanical behavior of such complex materials on the whole. Cases when macroscopic mechanical behavior of the polymer containing LM was primarily controlled by the form, in which this compound existed in polymer matrix, were considered. The future development of this direction of polymer modification was discussed and possibilities of its practical application for controlled improvement in the mechanical response of polymer materials were outlined.

Introduction

In many cases physical modification of polymer materials may be achieved by introduction of compatible low-molecular-mass additives (plasticizing agents, dyes, reinforcing agents, flame-retardants, etc.) into polymers. Compatibility of the components is controlled by nonchemical interaction between molecules of low-molecular-mass compounds (LMs) and polymer chains. Such interaction is provided by hydrogen bonding, electrostatic, dispersion, and similar interactions, which, in the general case, are responsible for complexation between the components of polymer composites. Physical and mechanical characteristics of the as-modified polymer materi-

sulfone (II), synthetic tanning agent BNS (III) and resorcinol-formaldehyde novolac (IV) were used as OACs.

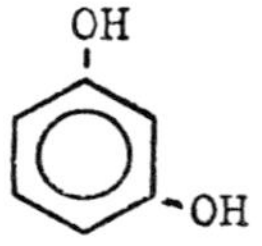

I

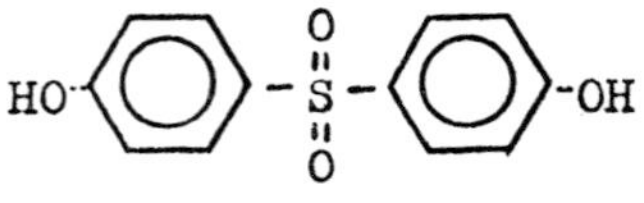

II

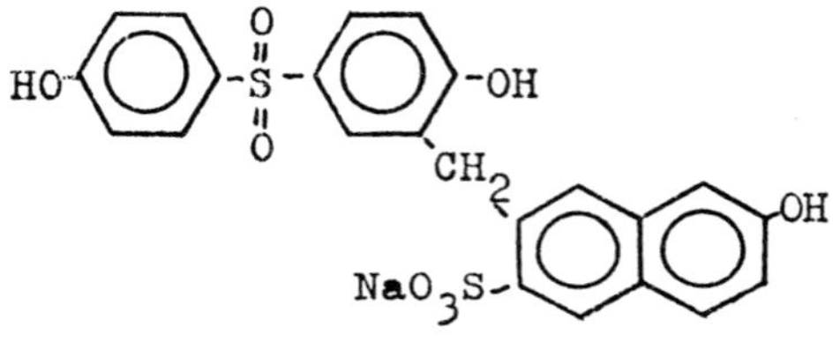

III

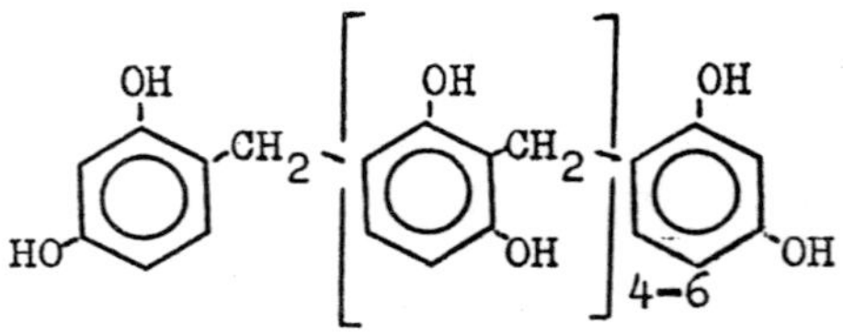

IV

As was shown, only compounds III and IV are able to produce IB form, and compounds I and II are not. The studies on sorption of water vapors by the PA-6/OAC samples at 20°C allowed one [2] to conclude that IB form of compound III is primarily localized in the densely packed regions of the amorphous phase of PA-6, which are unaccessible for water molecules at sorption temperatures in a whole range of relative humidities. The RB form occupies disordered regions of amorphous phase, the water penetration into which is known to control the water sorption by the polymer samples [1, 20].

In the case of compound IV, the formation of IB form takes place in the loosely packed regions of the amorphous phase of PA-6. However, there is still an open question: Do molecules of compound IV penetrate the densely packed domains of amorphous phase in PA-6 or they are localized only in the loosely packed regions?

From the standpoint of complexation, the distribution of LM between two forms may be represented as a result of occurrence of two reactions of complexation between the components of the PA-6/OAC system:

$$[OAC] + [CONH] \rightleftarrows [OAC \cdots CONH] \quad (1)$$

$$[OAC] + [CONH] \rightarrow [OAC \cdots CONH] \quad (2)$$

The difference between the above reactions is related to reversible character of reaction (1) and irreversible character of reaction (2). The subsequent treatment of the PA-6/OAC samples with a pure solvent (water) is associated with shifting the equilibrium state of reaction (1) to the left until complete dissociation of the complex. At the same time, the complex formed via reaction (2) shows no dissociation under the above conditions.

Examination of published evidence on complexation of PET [21] and poly(ethylene oxide) (PEO) [22] with resorcinol in melt, poly (N, N-dimethylacrylamide) with phenol-formaldehyde resins in the solutions [23] and the general speculations concerning the mechanism of complexation between polymer and oligomer chains [24] allowed one to specify two interdependent conditions for the development of stable intermolecular complexes:

1. Realization of joint dense packing of the molecules of the reacting compounds [21, 22]. In other words, reacting components should be complementary: geometric structure of the interacting species should not prevent the development of intermolecular bonds [24];

2. The complexation reaction should involve a certain sequence of active complexing groups, the number of which is equal to or exceeds crucial one [23, 25]. This condition implies cooperativity of a set of intermolecular bonds in the complexes with polymer chains. Just cooperativity of interaction is primarily responsible for relatively high stability of the reaction products as compared with that of the complexes between the low-molecular-mass analogs.

Let us first discuss what do different authors mean when speaking about a "stable complex."

For example in [24], "stable complex" is treated from the kinetic standpoint. In the case of complexation reactions between polymer chains and oligomers, an increased stability of the formed complex was accounted for by an increase in the constant of equilibrium, i.e. by shifting the equilibrium towards the complex formation. The above reactions were shown to possess reversible character. In particular, the reaction of complexation between polymer and oligomer chains is reversible under varying the concentration of oligomer in the solution: a decrease in concentration of the oligomer is accompanied by shifting the equilibrium towards accumulation of the initial reagents until complete dissociation of the complex.

On the other hand, in the case of the complexation between poly (N, N-dimethylacrylamide) and phenol-formaldehyde resins [23] in organic solvents, the stable complex was referred to as the reaction product, which precipitated under the reaction conditions. Let us emphasize that the formed complex showed no dissociation under washing with a pure solvent. This evidence suggested irreversible character of the complexation reaction.

The similar situation was observed in the case of the complexation between aliphatic PA with OACs in the solutions of formic acid [6, 26] and alcohol [26]. In this case, compound III and sulfonated phenol-formaldehyde oligomer were used as OACs. Interaction of the above compounds with polymers (in particular, with PA-6) is accompanied by precipitation of the complex, which shows dissociation neither in a pure solvent nor in a boiling water.

A certain inconsistency in treating the term of "stable complex" does not conflict with the above conditions for complexation, which were advanced to explain the mechanism of specific interaction of OACs with polymer chains in the amorphous regions of PA-6. However, the prevent any terminological misunderstanding in this work "stable complex" is the complex, which shows no dissociation under treating in a pure solvent, i.e., which is formed via reaction (2).

The occurrence of reaction (2) in amorphous regions of PA-6 is allowed when the number of active complexing groups in OAC molecules is high enough, and such groups are able to react with polymer amide groups so that the number of reacting groups should be equal to or higher than critical one. The latter requirement is especially important for the complexation reactions in solid bodies, since as compared with solutions and melts, the macromolecular mobility in the solid phase is substantially inhibited (in our case, in amorphous regions of PA-6 swollen in aqueous solutions of OAC).

Inhibited mobility of polymer chains prevents the possibility for phenol hydroxyls of OACs and polymer amide groups to approach each other within the distance necessary for the development of hydrogen bonds. Furthermore, in the amorphous regions of PA-6, as compared with solution or melt, the mutual arrangement of macromolecules and, hence, the chain fragments with amide groups is fixed. Mutual arrangement of polymer amide groups is primarily controlled by polymer packing density in amorphous regions of PA-6. The heterogeneity in packing density in PA-6 was discussed in [18].

From this standpoint, the occurrence of reaction (2), i.e., the development of IB form, in the amorphous phase of PA-6 takes place only in the local structural regions which serves as a specific "trap" for OAC molecules. Just within such a trap, the above conditions for the development of stable complex are fulfilled and irreversible binding of OAC with polymer matrix is promoted. Let us discuss the experimental evidence obtained for complexation of PA-6 with OACs from this standpoint.

The development of only RB form for compounds I and II suggests that the chemical structure of the molecules of the above compounds (molecular sizes and number of the complexing groups) prevents the fulfillment of the necessary condition for the formation of stable intermolecular complex via reaction (2). In this case, interaction of phenol hydroxyls with polymer amide groups is described by reaction (1) and reversible character of binding of LM with polymer is realized.

When passing to OAC III, increase in molecular sizes and number of complexing hydroxyl groups result in appearance of IB form of LM. This implies that the number of active complexing groups is sufficient to produce stable complex with PA-6. However, the formation of IB form of OAC III takes place only within the densely packed regions of the polymer amorphous phase. This experimental evidence allows one to conclude that macromolecular arrangement only in the densely packed regions provides the conditions of cooperative interaction of all complexing groups of OAC with amide groups of PA-6, occurrence of reaction (2) and irreversible binding of OAC molecules with polymer matrix. In other words, the chemical structure of compound II molecules is so that the "traps" for them exist only in the densely packed regions of amorphous phase of PA-6. In the case of more loosely packed regions, the mutual arrangement of molecules in such regions prevents simultaneous interaction of amide groups with phenol hydroxyls of compound III in quantity necessary for the occurrence of reaction (2). As a result, the interaction between the components of the PA-6/OAC system is described by reaction (1) and the formation of only RB form is observed.

With further increasing the molecular sizes and the number of the active complexing groups, i.e., compound IV, one can observe the development of IB form also (or only) in the loosely packed regions of the amorphous phase of PA-6. This evidence suggests that, as compared with molecules of compound III, the molecules with such molecular structure are able to find the "traps" in the regions with lowered packing density.

The above speculations allow us to generalize the mechanism of distribution of LMs between IR and RB forms as advanced in [3].

Sorption by polymer of LMs from their aqueous solutions involves two stages. The first stage is associated with diffusion of molecules of LMs into the swollen regions of the polymer amorphous phase, which are characterized by lowered packing density. The second stage is related to diffusion of molecules of LMs from loosely to densely packed regions. The sorption is accompanied by binding of molecules of LMs by polymer chains both in loosely and densely packed regions of the polymer amorphous phase via complexation between the components.

In general case, complexation between the molecules of LMs and polymer chains in polymer amorphous phase can be described by reactions (1) and (2). The formation of IB form of LMs is described by reaction (2). The occurrence of this reaction is provided by fulfillment of the above conditions of development of stable intermolecular complex. The possibility of fulfillment of these conditions in the polymer amorphous phase is decided by the chemical structure of molecules of LMs (molecular size and functionality) and packing density of polymer chains. In connection with this, IB form of LMs may be realized either in loosely or in densely packed regions of polymer amorphous phase or does not form at all.

However, the mechanism of diffusion of molecules of LMs into the densely packed regions and their withdrawal from such regions is still unclear. The penetration of molecules of LMs into the densely packed regions proceeds from their aqueous solutions, which are able to swell polymer matrix. On the other hand, desorption of molecules of LMs from the densely packed regions implies transition of molecules of LMs from polymer matrix into the pure solvent - water. The question arises: Does water swell the above regions or, on other words, whether such regions are accessible to water molecules?

The data reported in [3] clearly demonstrated that the densely packed regions of amorphous phase of PAN are unaccessible for water molecules under sorption and desorption of Br_2 from aqueous solutions. The formation of IB form of dioxydine in copolymer of N-vinylpyrrolidone with methyl methacrylate was also treated [4] within the standpoint of the existence of structural zones in the amorphous regions of moderately hydrophilic polymers, which are unaccessible for water molecules. However, in the case of the PA-6/OAC systems such conclusions do not seem quite obvious.

As was shown in [2, 19], the densely packed regions of the amorphous phase of PA-6 are unaccessible for water molecules under water sorption from the vapor phase at 20°C in a whole range of relative humidities. On the other hand, water sorption of PA-6 was shown to increase by approximately 30% as the temperature increases from 25 to 100°C [27]. Hence, the assumption that densely packed regions of PA-6 are accessible to water molecules under sorption of OACs from aqueous solutions at 100°C and desorption on treatment the PA-6/OAC samples in boiling water seems to be quite reasonable.

In connection with this, let us discuss the importance of an exact knowledge whether the densely packed regions are accessible for water molecules with respect to the mechanism of distribution of LMs in the polymer between IB and RB forms. Furthermore, the solution of this problem seems to be rather interesting from the viewpoint that according to [27] the accessibility of amorphous regions of semicrystalline polymers (in particular, PA-6) for water molecules is not constant but depends on sorption temperature.

To gain a deeper insight into the mechanism of distribution of LMs in polymer matrix between two forms, let us consider both cases: densely packed amorphous regions are accessible for water molecules under sorption and desorption (i), and densely packed amorphous regions are unaccessible for water molecules under sorption and desorption (ii).

Let us first assume that the densely packed regions of the polymer amorphous phase are accessible for water molecules under sorption and desorption of LMs. The point is that this version agrees well with the above mechanism of distribution of LMs between IB and RB forms. Aqueous solutions of LMs swell the densely packed regions of polymer matrix and the molecules of LMs are able to penetrate these regions. The treatment of the polymer/LM systems in a pure solvent (water) is also associated with swelling of the above regions in water, however, taking into account the above reasoning, in some cases, no dissociation of the polymer/LM complex takes place.

According to the assumption that densely packed regions are unaccessible for water molecules, the amorphous polymer phase under sorption and desorption of LMs may be visualized as heavily swollen loosely packed regions and "dry" densely packed regions. In this case, diffusion of molecules of LMs into the densely packed regions is carried out through a sharp boundary between swollen and "dry" polymer. Actually, we deal with distribution of LMs between loosely packed water-swollen regions and "dry" densely packed domains. In the case of the PAN/Br_2 and PA-6/OAC systems, this explanation seems to be quite plausible due to a high affinity between polymer and LMs [3, 19].

On further treatment the polymer/LM samples in water, desorption of LMs from the densely packed regions proceeds as distribution of LMs between loosely paced water-swollen regions and densely packed regions containing this LM. The LM may be completely extracted by solvent from the densely packed regions only in the case when molecules of this agent show sufficient mobility within such structural units. Such situation is likely to be realized for the PA-6/OAC systems only for the compounds I and II, which are able to produce only RB form. The molecules of compound III appear to be immobilized only in the densely packed regions, and no complete extraction of this agent is allowed.

Hence, there is no conflict between the above speculations and the mechanism of distribution of LMs in the polymer matrix between RB and IB forms. In other words, this mechanism provides a plausible explanation of the formation of IB form of LM in polymer matrix independently of the fact whether densely packed regions are accessible for water molecules or not.

So, the development of IB form of LMs in polymer matrix is associated with peculiar features of complexation between molecules of LMs and polymer chains as a result of structural heterogeneity of the amorphous polymer state. In connection with this, the systems, which show the formation of IB forms only in the densely packed regions of the polymer amorphous phase are of most interest, that is, PAN/Br_2 [3] an PA-6/(OAC III) [2,5,19] systems. The point is that in this case we can realize a unique possibility of separate modification of different structural levels in polymer amorphous phase. Actually, sorption of LMs from their aqueous solutions by polymers is associated with penetration of molecules of LMs into loosely and densely packed regions of polymer amorphous phase. In other words, both structural levels appear to be modified. The removal of RB fraction of LMs from the loosely packed regions allows one to obtain the polymer samples, in which only one structural level with dense packing of polymer chains is modified.

To understand the peculiar features and possibilities of such separate modification of structural levels of the polymer amorphous phase, let us carry out a comparative analysis of experimental evidence obtained for the polymer systems based on PA-6 and OAC III.

STRUCTURAL FEATURES OF MECHANICAL AND PHYSICAL BEHAVIOR OF POLYMER/ (LOW-MOLECULAR-MASS COMPOUND) SYSTEM

The systems based on PA-6 and OAC III are of great interest from the viewpoint of realization of separate modification of different structural levels in the polymer amorphous phase [2, 19]. The treatment of PA-6 films in aqueous solutions of OAC III is associated with modification of both structural levels (Samples I). The removal of RB form from the loosely packed regions implies that only one structural level, that is, densely packed regions, appears to be modified (Samples II).

The temperature dependences of storage modulus E' of dry samples of initial PA-6 samples and Samples I and II (Figure) allows one to outline the following general features of physical and mechanical behavior of such systems [5].

First, the temperature dependences are characterized by the existence of the inversion point of elastic properties of the materials, T_I. At temperatures below T_I, OAC III shows reinforcing action on the polymer elastic properties, whereas at temperatures above TI this action is weakening.

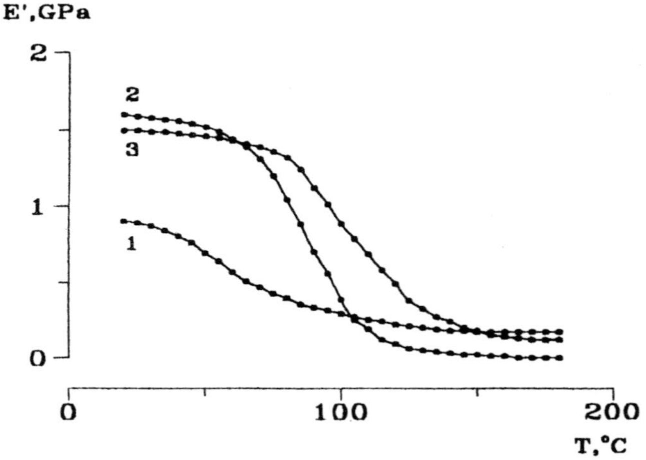

Figure 1. Temperature Dependence of Storage Modulus E' for PA-6 (curve 1) and PA-6/(OAC III) Samples. OAC III Content: 1.1 mol/kg (Sample I, curve 2), and 0.31 mol/kg (Sample II, curve 3).

Second, the removal of RB form from PA-6 (transition from curve 2 to curve 3) is accompanied by shifting T_I towards higher temperatures.

The above behavior was described in [5] in terms of polymer antiplasticization-plasticization with LMs [28, 29]. Molecular mobility of LMs (for example, OAC III) in polymer increases with temperature. An increased molecular mobility of LMs is associated with a decrease in energy of intermolecular interaction between polymer chains, and this provides the conditions for the transition from reinforcing (antiplasticizing) action to weakening (plasticizing) action of OAC III on PA-6. The macroscopic manifestation of this behavior is associated with the appearance of inversion point T_I in the corresponding temperature dependence of elastic properties of polymer systems.

Examination of the behavior of elastic properties of the PA-6/(OAC III) systems in a temperature range from 100-110 to 14-150°C is of particular interest. In the case of Samples I containing OAC III in both IB and RB forms, the corresponding elastic modulus E' is lower than that of initial polymer samples (curve 2) in this temperature range. For samples II containing only IB form of LMs, the value of E' appear to be higher than those of initial PA-6 samples.

From the viewpoint of molecular dynamics, this experimental evidence suggest that in the above temperature range the molecular mobility of RB form is well-pronounced. Consequently, the RB molecules are able to plasticize the polymer, and a concomitant decrease in elastic modulus is observed. In contrary, molecular mobility of IB form is still frozen-in and its action on polymer matrix is antiplasticizing (reinforcing). In other words, the mobility of RB fraction of OAC III appears at temperatures by 40-50°C lower as compared with IB form.

According to our theoretical speculations the difference between the molecular mobilities of RB and IB forms in a certain temperature range may be rationalized by the fact that IB form is localized in the densely packed domains of the amorphous phase, which are characterized by relatively high level of intermolecular interaction. In connection with this, manifestation of molecular mobility of IB form of OAC is allowed only at elevated temperatures. As the RB fraction of OAC occupy only loosely packed regions of the amorphous phase, which are characterized by lower energetics of intermolecular interaction, then the molecules of this RB fraction acquire mobility at much lower temperatures.

The above evidence allowed us to draw the following conclusions concerning the specific character of elastic behavior of the PA-6/(OAC III) samples. In the aforementioned temperature interval, the situations is realized when one and the same modifying agent is able to have quite different (even opposite) effect on the elastic properties of the material, depending on the form of its existence within the polymer matrix. In other words, in a given temperature range, elastic properties of the PA-6/(OAC III) samples containing OAC in both RB and IB forms are controlled by interplay between plasticizing action of RB form and antiplasticizing action of IB form. The net result depends on the ratio between IB and RB forms of OAC in polymer matrix [5]. This ratio may be changed either by varying the concentration-time regimes of introduction of OAC into the polymer matrix or by desorption of RB fraction of LM from PA-6 [2].

Taking into account the above speculations, one may easily explain the changes in glass transition temperatures T_g of PA-6 samples on sorption and desorption of OAC III (Table 1) [5]. Introduction of OAC into the polymer matrix is accompanied by increase in glass transition temperature T_g of the polymer (Samples I, Table 1). This is main difference between the systems studied and traditional plasticized polymer materials [28, 29]. The removal of RB form of OAC III from the test samples, which is known to exert plasticizing action on the polymer at temperatures above 100°C, i.e., transition from Samples I to Samples II, is accompanied by a growth in T_g of PA-6 samples (Table 1).

So, examination of physical and mechanical behavior of the PA-6/(OAC III) samples together with speculations concerning the structural heterogeneity of the polymer amorphous phase may serve as a convincing demonstration of possibilities of controlled modification of the bulk properties of polymer materials via controlled modification of different structural levels of amorphous phase of semicrystalline polymer. This conclusion is also supported by comparing the properties of Samples I and II in humid atmosphere (Table 2).

Table 1 . Glass transition temperatures of dry PA-6/(OAC III) Samples.

Sample	Content of OAC III in PA-6, mol/kg	Tg, °C
Initial PA-6	0	80
Sample I	0.52	115
Sample II	0.15	123
Sample I	1.1	120
Sample II	0.28	133

In the case of aliphatic polyamides, sorption of water vapors is known to be accompanied by decrease in elastic modulus of the polymer samples [30, 31]. As compared with initial PA-6, the PA-6/(OAC III) samples containing both IB and RB forms of OAC III (Sample I, Table 2) are characterized by a sharp decrease in water sorption and increase in elastic modulus [2]. Desorption of RB fraction of OAC III from PA-6 (transition from Samples I to Samples II, Table 2) is accompanied by an abrupt increase in water sorption, which is comparable to that of initial polymer. However, no increase in elastic modulus is observed. In other words, distribution of OAC III between IB and RB forms is associated with peculiar "division of labor": RB fraction of modifying agent occupies loosely packed structural levels of polymer amorphous phase and it responsible for water sorption of the polymer material, whereas macroscopic elastic properties of the materials in humid atmosphere are controlled by IB fraction localized in the densely packed regions. Furthermore, such complex modification results in reduced susceptibility of the resultant materials to the plasticizing action of water as evidenced by constancy of elastic modulus even at two-fold increase in water sorption when passing from sample I to sample II (Table 2).

Table 2. Elastic properties of the PA-6/(OAC III) samples in humid atmosphere.

Sample	Content of OAC III in PA-6, mol/kg	p/o_0, %	Water content, mol/kg of amorphous phase	Elastic modulus, GPa
Initial PA-6	0	0	0	1.9
	0	47	2.8	1.3
Sample I	0.76	47	0.95	2.5
Sample II	0.23	47	2.4	2.5

The above evidence on the effect of IB fraction of OAC III localized in the densely packed domains of amorphous phase on macroscopic elastic properties of the polymer samples allows one to draw an important conclusion concerning a marked role of such domains in development of mechanical properties of semicrystalline polymers.

CONCLUSION

Experimental evidence discussed in this work unequivocally shows that in some cases a complicated character of interaction of LMs with polymers is realized. The LMs introduced into the polymer matrix are shown to be distributed between two different forms: reversible (RB) and irreversibly bound (IB) forms. All systems, which show similar behavior, are characterized by high affinity between LMs and polymers, which is controlled by complexation between molecules of LM and polymer chains.

The mechanism of distribution of LMs between IB and RB forms is based on speculations concerning structural inhomogeneity of amorphous component in semicrystalline or completely amorphous polymers: the existence of noncrystalline regions (structural levels) with different packing density and order. The development of IB form of LMs is a result of complexation reaction between molecules of LM and polymer chains and formation of very stable complex. The possibility of occurrence of such complexation reactions is dictated by chemical structure of LM and local packing density of polymer chains. In connection with this, IB form may be localized within different structural levels or even does not form at all.

In turn, the complicated character of interaction between LMs and polymer chains is responsible for a complicated molecular dynamics of polymer/LM systems and, hence, for specific features of mechanical behavior of the material.

The approaches discussed in this work may be interpreted as underlying principles for new scientific direction, according to which interaction between LMs and polymers is primarily treated from the viewpoint of existence of different levels of structural inhomogeneity in polymers. This direction involves some general problems related to sorption and diffusion of LMs in polymer bodies with a complicated structural organization.

On the other hand, controlled occupation of different levels of structural inhomogeneity with molecules of LM suggests many possibilities of controlled modification and improvement in mechanical properties of polymer materials.

Further development of this direction should involve solution of the following important problems:

1. The role of structural inhomogeneity (in other words, contribution of each structural level) in the development of physical and mechanical properties of polymer materials;
2. Selective modification of each structural level.

This direction may be conditionally referred to as "structural mechanics of polymers," which is associated with the effect of fine structures on physical and mechanical behavior of polymer materials.

REFERENCES

1. Arzhakov, M.S., Volkov, A.V., Volynskii, A.L., and Bakeev, N.F., Makromol. Chem., Macromol. Symp., 1991, vol. 44, p. 165.
2. Arzhakov, M.S., Red'ko, N.V., Volkov, A.V., Volynskii, A.L., and Bakeev, N.F., Polym. Sci., A, 1994, vol. 36 (in press).
3. Lewin, M., Guttman, H., and Naor, Y., J. Macromol. Sci., A, 1988, no. 10-11 p., 1367.
4. Polishchuk, A.Ya., Zimina, L.A., Madyuskin, N.N. and Zaikov, G.E., Polym. Sci., A, 1993, vol. 35, no 1., p. 80.
5. Arzhakov, M.S., Nechaeva, E.V., Volkov, A.V., Volynskii, A.L., and Bakeev, N.F., Polym. Sci., A, 1993, vol. 35, no. 4, p. 521.
6. Volkov, A.V., Arzhakov, M.S., Bondarev, V.V., Volynskii, A.L., and Bakeev, N.F., Vysokomol. Soedin, A, 1990, vol. 32, no. 4, p. 467 (in Russian).
7. Volkov, A.V., Arzhakov, M.S., Volynskii, A.L., and Bakeev, N.F., Dokl. Akad. Nauk SSSR, 1989, vol. 306, no. 6, p. 1401 (in Russian).
8. Guthrie, J. and Cook, C.C., Polymer, 1981, vol. 22, no. 10, p. 1439.
9. Kishimoto, Y., Hosaka, T., and Shimotsuma, W., J. Appl. Polym. Sci., 1986, vol. 32, p. 5161.
10. Topic M., Mogus-Milankovic A., and Katovic Z., Polymer, 1987, vol. 28, p. 33.
11. Volynskii, A.L., Volkov, A.V., Bondarev, V.V., Seleznev, N.I., Arzhakov, M.S. and Bakeev, N.F., Vysokomol. Soedin. A, 1990, vol. 32, no. 11, p. 2323 (in Russian).
12. Preovrsec, D.C., J. Polym. Sci., C, 1971, vol. 32, p. 343.
13. Prevorsec, D.C., Tirpac, G.A., Harget, P.J., and Reimschuessel, H.C., J. Macromol. Sci., B, 1974, vol. 9, p. 733.
14. Yeh, G.S.J., J. Macromol. Sci., B, 1972, vol. 6, p. 465.
15. Klement, I.J. and Geil, P.H., J. Macromol. Sci., B, 1971, vol. 5, p. 505.
16. Arzhakov, S.A., Bakeev, N.F. and Kabanov, V.A., Vysokomol. Soedin, A, 1973, vol. 15, no. 5, p. 1154 (in Russian).
17. Kozlov, G.V., Sanditov, D.S., and Serdyuk, V.D., Polym. Sci., A, 1993, vol. 35, no. 12, p. 1742.
18. Perepechko, I.I. and Yakovenko, S.S., Vysokomol. Soedin., A, 1981, vol. 23, no. 5, p. 1166 (in Russian).
19. Arzhakov, M.S., Nechaeva, E.V., Red'ko, N.V., Volkov, A.V., Volynskii, A.L., and Bakeev, N.F., Polym. Sci., A, 1994, vol. 36 (in press).
20. Volkov, A.V., Arzhakov, M.S., Volynskii, A.L., and Bakeev, N.F., Vysokomol. Soedin., A, 1990, vol. 32, no. 3, p. 489 (in Russian).

21. Myasnikova, R.M., Samarskaya, V.D., and Obolonkova, E.S., Vysokomol. Soedin., B, 1992, vol. 34, no. 2, p. 43 (in Russian).
22. Myasnikova, R.M., Titova, E.F., an Obolonkova, E.S., Polymer, 1980, vol. 21, p. 403.
23. Yang, T.P., Pearce, E.M., Kwei, T.K., and Yang, N.L., Macromolecules, 1989, vol. 22, p. 1813.
24. Kabanov, V.A. and Papisov, I.M., Vysokomol. Soedin., A, 1979, vol. 21, no. 2, p. 243 (in Russian).
25. Ilipoulos, I., and Audebert, R., J. Polym. Sci., B, 1988, vol. 26, p. 2093.
26. Arzhakov, M.S., Cand. Sci. (Chem.) Dissertation, Moscow: Lomonosov Moscow State University, 1989 (in Russian).
27. Razumovskii, L.P., Markin, V.S., and Zaikov, G.E., Vysokomol. Soedin, A, 1985, vol. 27, no. 4, p. 675 (in Russian).
28. Kozlov, P.V. ad Papkov, S.P., Fiziko-Khimicheskie Osnovy Plastifikatsii Polimerov (Physicochemical Foundations of Plasticization of Polymers), Moscow: Khimiya, 1982 (in Russian).
29. Nielsen, L.E., Mechanical Properties of Polymers and Composites, New York: Marcel Dekker, 1974.
30. Water in Polymers, Ed., Rowland, S.P., Washington, D.C., 1980.
31. Puffr, R., and Sebenda, J., J. Polym. Sci., C, 1967, no. 16, p. 79.

Matrix Polycondensation as a New Route to Multicomponent Polymer Systems with Prespecified Properties

A. Kuznezov, V. Vasnev, I. Gribova,
A. Krasnov, G. Gureeva, and V. Ignatov
Institute of Organo-Element Compounds
of the Russian Academy of Sciences,
Vavilov Str. 117334, Moscow, Russia

Abstract—**The regularities of polyesterification in the presence of a filler as a heterogeneous matrix were considered. The filler introduced into the reaction system before polycondensation process influenced not only the reaction kinetics but also the resulting properties of polymer composites produced. The composites obtained by polymer synthesis in the presence of the filler have higher mechanical, thermal, frictional and electrical properties as compared with those of the composites prepared by mechanical filling.**

Introduction

The term matrix reaction is usually used for polymerization processes in the other polymer which can affect the reaction kinetics, structure and properties of the daughter polymer. While the addition matrix polymerization was studied well, the matrix polycondensation (step-growth polymerization) is still not developed. The majority of experimental data was obtained in the synthesis of aromatic polyamides (1-3). Highashi et al (1-3) have established that some flexible chain polymers introduced into the reaction system increase the molecular weights of rigid-rod aromatic polyamides obtained by the phosphorilation reaction directly from acids and diamines. Authors (3) have found that in the presence of poly (vinylpyrrolidone) as a matrix it was possible to obtain high-molecular poly-(p-phenylene terephtalamide) soluble even in the organic solvent.

In our opinion, matrix polycondensation is one of the best methods to design a polymer multicomponent system (miscible blends, molecular composites, composites, and etc.) with prespecified properties. However, the role of the matrix can play not only the polymers dissolved in the reaction system but the filler particles dispersed within the reaction volume. In this case it is possible to regulate the structure and the properties of the polymer synthesized by varying the content of the filler and its specific surface area.

In our previous work (4) we have discovered the heterogeneous matrix effect of antifriction filler (graphite and molybdenum disulfide) in acceptor-catalytic polyesterification. It was established that the filler selectively adsorbed monomers and, as a consequence, influenced their reactivity. For instance, the reactivity of the aromatic hydroxyl groups after sorption was lowered almost 3 times. However, the molecular weights of polymers obtained in the presence of 30 to 60% filler were more that two times higher as compared with those of the polymer obtained without the filler.

In this work the effect of the filler on the properties of polymer composites based on polyterephtalate of bis (4-oxy-3-chlorophenyl)-2,2'-propane and graphite is considered in more detail.

Materials and Methods

The substances and solvent were purified using standard methods. Polyterephtalate of bis (4-oxy-3-chlorophenyl)-2,2'-propane (dichlorodiane) was synthesized by acceptor-catalytic polyesterification in the presence of a filler amounting to 30-60% of the calculated weight of the filled polymer composite (4).

It was established that the polyesterification process occurred under quasihomogenous conditions in the kinetic range.

The molecular weights of the polymers were determinated by light scattering after separation of the filler by filtration followed by centrifugation of the composite solution in chloroform.

Microphotographs of the surface of the coatings based on the composites obtained directly during the polymer synthesis and by mechanical filling were made on a Hitachi scanning microscope.

The thermal, frictional, and mechanical properties and electrical conductivity of the composites were measured by standard techniques.

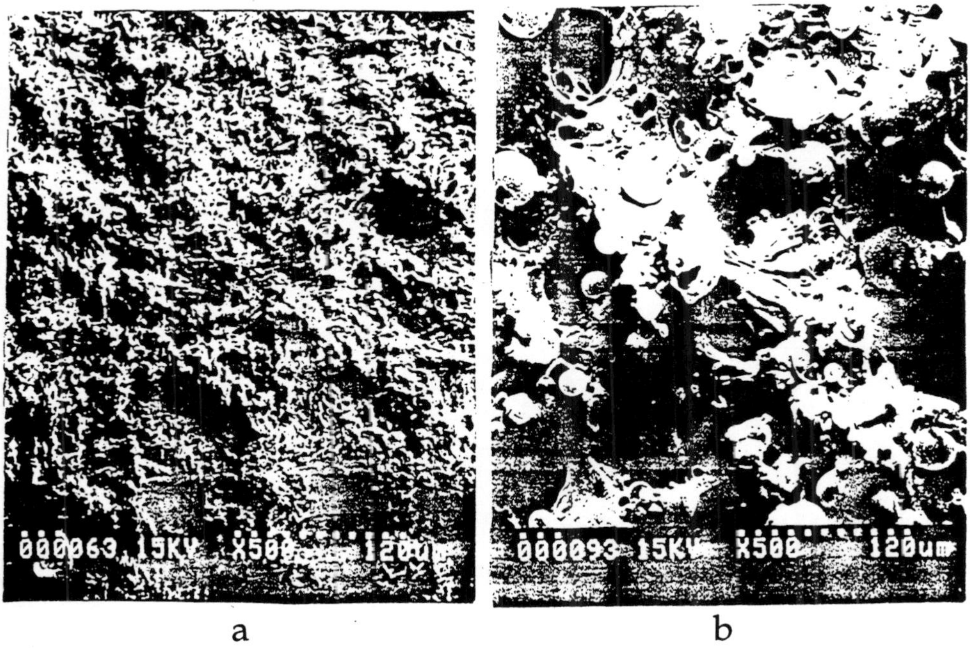

Figure 1. Microphotographs of surfaces of the coatings of a composite based on the dichlorodiane polyterephtalate and 30% graphite S-1: a) synthetic composite; b) mechanically filled composite.

Results and Discussion

The properties of polymer composites used, depend on many factors, first of all, on the molecular weight of polymer, its structure, homogeneity of filler distribution within the polymer phase, orientation of polymer chains on the filler surface, interaction between filler particles and macromolecules, and etc.

The results presented in the Table 1 reveal a possibility to regulate the molecular weight of polymers by content of the filler introduced into the reaction system before the synthesis. As we have established (4) this effect is caused by selective sorption of the polarized intermediate active complexes of the monomers on the filler surface.

Table 1. Effect of a filler on the molecular weight of dichlorodiane poly-terephtalate (Benzene, Et_3N, 30°).

Filler, (%)		Specific surface, S_{sp}, m^2/g	[n], 100 mL/g (tetrachloroethane, 25°)	Molecular weight, $M_w \cdot 10^3$
No Filler		–	0.68	30
Graphite "Taiginskii"	(30)	9.5	0.87	53
Graphite S-1	(30)	14.0	0.92	60
Graphite S-1	(60)	14.0	1.15	70

Because of the matrix effect a growth of macromolecular chain occurs at the surface or in the region close to the surface of the heterogeneous matrix. For this reason the amtrix effect provides an opportunity to produce polymers with higher molecular weights.

The microphotographs of the coatings based on the synthetic and mechanically filled composites demonstrate more homogeneous distribution of the filler particles on the first case (Figure 1a). The coatings based on the mechanical composite are more enriched with polymer composite solution on the metal plate. This means that in the mechanically filled composite the interaction between polymer molecules themselves is dominated over the interaction polymer-filler.

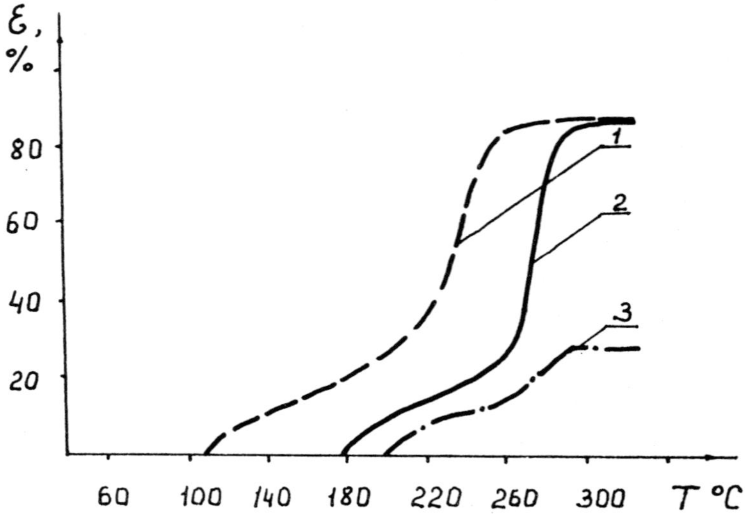

Figure 2. Data of thermomechanical tests of the composites: 1-mechanically filled composite; 2,3-synthetic composites with 30% and 60% graphite S-1, respectively.

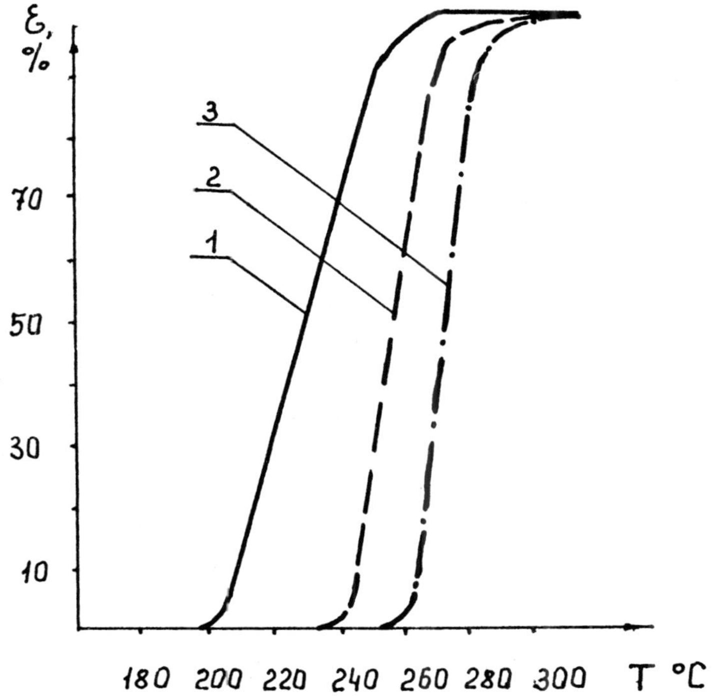

Figure 3. Data of thermomechanical tests of polymers (table 1) obtained without graphite (curve 1) and in the presence of 30% (curve 2) and 60% graphite S-1 (curve 3).

For this reason the mechanical composite disintegrates because of adhesive destruction of the bonds between polymer and filler, while the synthetic composite fractures because of cohesive destruction of the polymer phase itself.

The results of thermomechanical tests reveal an increased heat resistance of synthetic composites (Figure 2). The softening temperature of the synthetic composite with 30% of graphite is 70° higher as compared with the mechanically filled system. For synthetic composite with 60% of graphite this value increases up to 90°. One of the reasons could be an increased molecular weight of polymers in the synthetic composites (Table 1).

Polymers obtained in the presence of graphite have an increased heat resistance (Figure 3). However, the difference in the softening temperatures is lower (40°). The greater difference in softening temperatures of the composites could be explained by a more intense interaction between polymer molecules and filler particles in the synthetic composite.

The thermomechanical tests of the synthetic and mechanical composites with 30% of graphite S-1 and polymers with the same molecular weights (50

000) confirm the conclusion about more intense interaction between polymer and filler in the synthetic composite. The softening temperature in the latter case is much higher than for the mechanical composite (Figure 4).

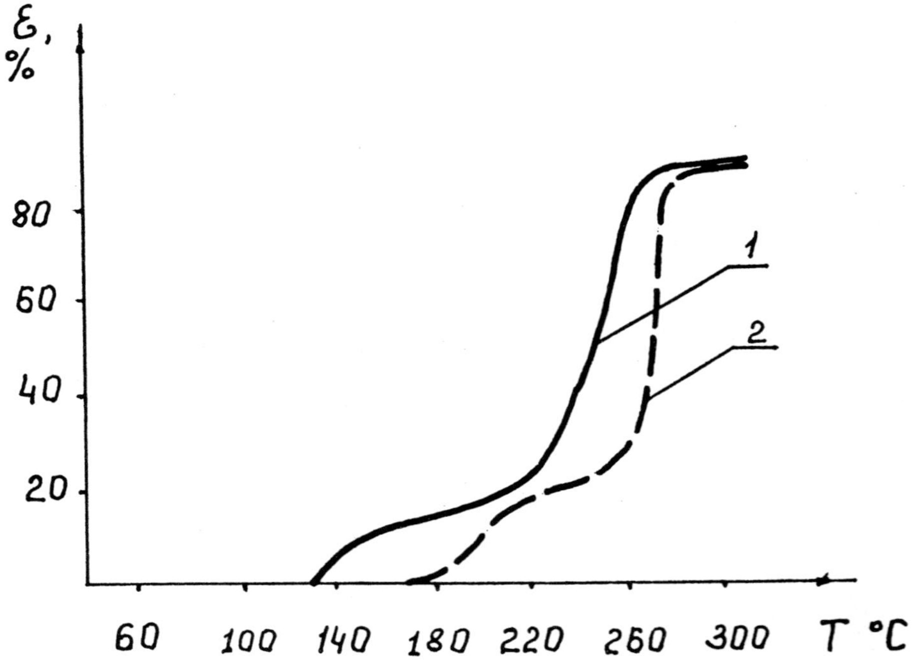

Figure 4. Thermechanical curves of the composites based on polyterephtalate of dichlordiane (M_W=5.10^4) with 30% graphite S-1: 1-mechanically filled composite; 2-synthetic composite.

The method of filling influences also the thermal stability of the composites. The thermal degradation of composites by destruction of the ester groups at 350° in vacuum appeared to be much higher in the case of the mechanically filled composite. According the results of chromatography study of gaseous products of thermal and dioxide evolved from the mechanically filled composite are 10 and 30 times higher, respectively, than those of the synthetic composite (Table 2). This is also caused by a more intense interaction between polymer and filler in the synthetic composite.

Table 2. Properties of composites based on dichlorodiane polyterephtalate and graphite S-1.

Type of composite	Thermolysis $CO\ 10^3$ mol/unit-mol	$CO_2\ 10^3$ mol/unit-mol	Specific impact strength, kJ/m^2	Specific electroconductivity $\varepsilon.10^{13}\ ^{-1},\ m^{-1}$
Mechanical composite	9	63	45	1
Synthetic composite	1	2	118	500

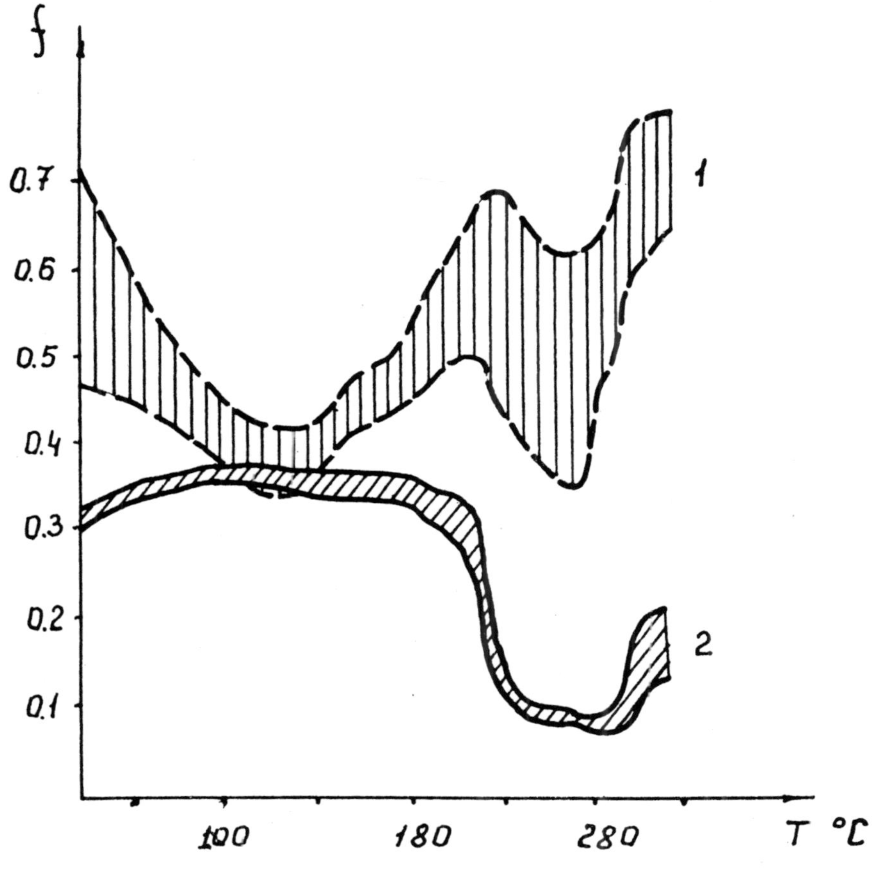

Figure 5. Thermofrictional properties of coatings based on the dichlorodiane polyterephtalate and 30% graphite S-1: 1-mechanical composite; 2-synthetic composite.

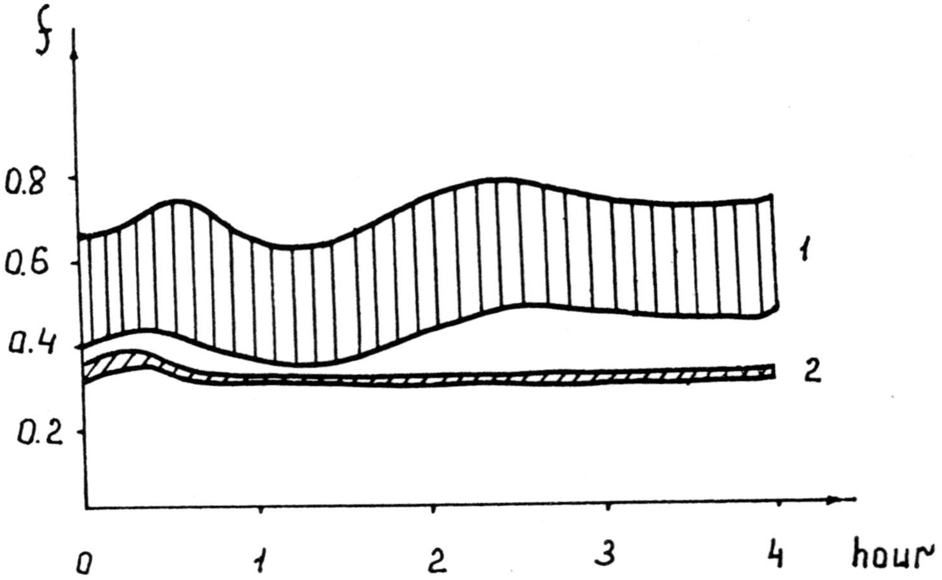

Figure 6. Frictional properties of coatings based on the dichlorodiane polyterephtalate and 30% graphite S-1: 1-mechanical composite; 2-synthetic composite.

The data presented in Table 2 reveal another advantage of synthetic filling. For instance, the value of specific impact strength of the synthetic composite is 2.5 times higher than that of the mechanical one. The electrical conductivity of the synthetic composite is 500 times higher.

The friction properties of the composites also demonstrate the advantage of synthetic filling. The data presented in Figures 5, 6 and 7 show a more stable character of friction of coating based on the synthetic composite. The friction coefficient of this coating is three times lower and it tends to decrease with an increase of the friction temperature (Figure 5) and test duration (Figure 6), while the friction coefficient of the coating based on the mechanical composite increases with an increase of temperature and test duration. Even 80% of graphite in the composites does not change the unstable character of friction of the mechanical composite (Figure 7).

This can be caused by nonhomogeneous distribution of a filler in the mechanical composite. In an accordance with the electron microscope data the surface of the coating based on the mechanical composite is more enriched with polymer. This fact leads to more intensive thermoxydation of the polymer on the frictional surface and, as a consequence, to an increase in the friction coefficient.

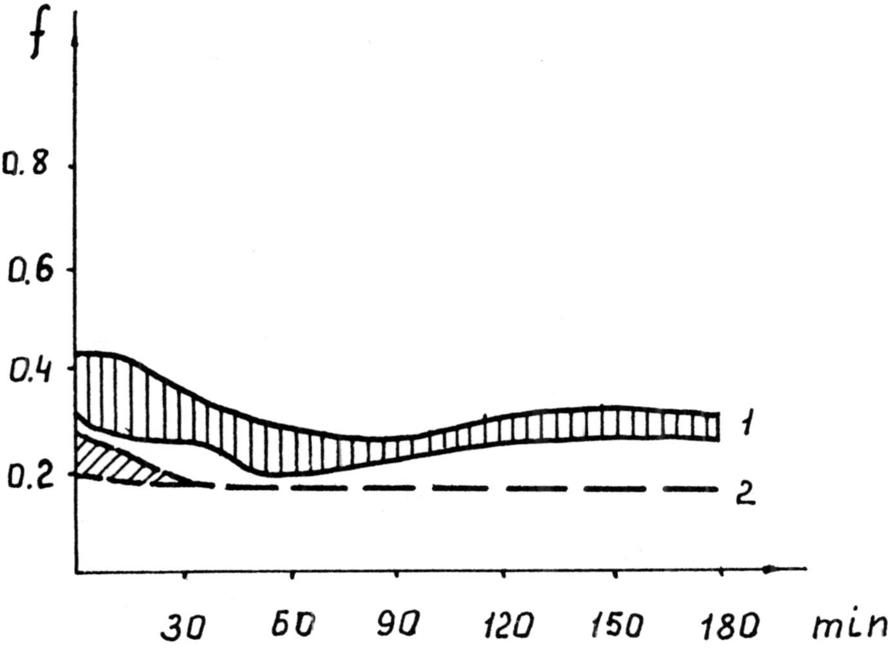

Figure 7. Frictional properties of the coatings based on the dichlorodiane polyterephtalate and 80% graphite S-1: 1-mechanical composite; 2-synthetic composite.

These results are confirmed by values of the free surface energy, γ_s, determined by a standard technique (5) (Table 3).

Table 3. Dependence of free surface energy $\gamma_s=\gamma_h+\gamma_d$* on the duration of friction test of the coating based on dichlorodiane polyterephtalate and 30% graphite S-1.

Duration of friction test, hour	Mechanical $\gamma_h \cdot 10^3$, J/m^2	composite $\gamma_d \cdot 10^3$ J/m^2	Synthetic $\gamma_h \cdot 10^3$, J/m^2	composite $\gamma_d \cdot 10^3$ J/m^2
Before friction	2.5	50.2	1.2	58.0
1	2.9	47.4	–	–
2	10.3	41.7	1.1	58.1
3	19.2	41.0	0.9	58.0

*γ_h and γ_d are the polar and dispersional parts of free energy γ_s.

The data presented in the Table 3 show that the polar part of free energy, γ_h, considerably increases during friction of the coating based on the mechanical composite while in the case of the synthetic composite it becomes constant. This indicates the higher thermoxidative resistance of the surface of synthetic composite. The greater value of dispersion part of free energy, γ_d, in the case of the synthetic composite can be caused by its more ordered structure as compared with that of mechanically filled composite.

Summarizing all mentioned above we can conclude that the synthesis of polymers in the presence of heterogeneous and homogeneous matrices (fillers, polymers) provides the possibility to create the polymer multicomponent systems with the properties which can be regulated by the content and nature of matrices themselves.

REFERENCES

1. N. Yamazaki and F. Higashi. *Bull. Chem. Soc. Japan,* 1973, 46, 1235.
2. N. Yamazaki, M. Matsumoto and F. Higashi. *J. Polym. Sci., Polym. Chem. Ed.,* 1975, 13, 1372.
3. N. Yamazaki and F. Higashi, *Adv. Polym. Sci.,* 1981, 38, 1.
4. V. Korshak, V. Vasnev, I. Gribova et al., *Vysokomolec. Soed.,* 1989, 31 (A), 86-90.
5. D. Van Krevelen, *"Properties of Polymers correlation with chemical structure," "Khimia,"* Moscow, (1989), s. 567.

Physical Chemistry of Topological Disorders in Polymeric Substances. Rearrangement of Additive Sorption Centers

Yu.A. Shlyapnikov, N.N. Kolesnikova
Institute of Chemical Physics, 117334, Moscow, Russia

Abstract—Low-molecular compounds, if their concentrations in the melt of isotactic polypropylene exceed a certain limit, transform some elements of disorder into additional sorption centers for the same compounds. These centers can exist after removal of the low-molecular compound, which may be considered as an effect of post-plasticizing, but slowly disappear at elevated temperatures (80-100°C). The theory of sorption centers rearrangement is discussed.

Theory

In studying some physico-chemical processes in a polymer matrix, the problem of quantitative description of the polymer structure arises. To solve this problem, a two-component model of polymeric substance has been proposed by one of the authors [1-3]. According to this model, a bulk polymer is considered as a two-component mixture, the main component of which is the system of ordered polymer chains, and the second - the multitude of various units of disorder (UODs): knots, folds and various entanglements of the polymer chains. Because these structures are connected to long polymer chains they are stable even in a polymer melt where they behave as relatively mobile quasi-particles. UODs prevent complete crystallization of crystalline polymers and participate in some other processes. Being the sites of decreased density, they are the stable elements of free volume present in the polymer.

Despite the great variety of UODs, in respect of any process, they may be divided into two groups: those participating and those non-participating.

The molecules of a compound dissolved in the polymer need the elements of free volume to be present in. These elements may either pre-exist in the polymer or be formed by penetrating molecules. In this study we tried to find the relation between these two ways of low-molecular compound dissolution.

As accepted in dual mode sorption model [4], the additive A dissolution in a polymer includes formation of true solutions obeying the Henry law:

$$[A]=\gamma[A]_m \tag{1}$$

followed with sorption by sorption centers, which process may be considered as complex formation:

$$A+Z_i \underset{}{\overset{K_{ai}}{\rightleftarrows}} AZ_i \tag{2}$$

Here [A] is the concentration of truly dissolved A, and $[A]_m$ - of the same additive in surrounding medium (in vapor phase or in low-molecular solvent). According to two-component model, these centers are formed by some UODs containing enough space for A molecules.

It follows from (2) that the total additive concentration in the polymer $[A]_p=[A]+\Sigma[AZ_i]$ or [3,5]:

$$[A]_p=[A]+\sum_i \frac{K_{ai}\gamma_a[Z_i][A]_m}{1+K_{ai}\gamma_a[A]_m} \tag{4}$$

It has been shown in [5] that if $[A]<<\Sigma[AZ_i]$ for any symmetrical distribution of sorption centers concentration $[Z_i]$ on the heat of sorption, the isotherm (5) is virtually indistinguishable from Langmuir-type one and we may substitute the formula (5) with more simple one:

$$[A]_p+\sum_i \frac{K_a\gamma_a[Z_a][A]_m}{1+K_a\gamma_a[A]_m} \tag{5}$$

Where $[Z_a]=\Sigma[Z_i]$ for all centers participating in A sorption, and K_a is a certain average meaning of K_{ai}.

Numerous experiments had shown that the concentration of centers participating in sorption of any additive $A[Z_a]$ does not depend on temperature in wide intervals of the latter, sometimes exceeding 180°C [5].

If the concentration of A in the medium surrounding the polymer $[A]_m$ is in equilibrium with excess of pure A, $[A]_p$ will be equal to S_A - the additive solubility in it. It has been shown in [6] that, unlike the solubility in non-mixing liquids, S_A depends on the composition of medium surrounding the polymer because the molecules of solvent compete with the additive for some sorption centers ($AZ_i+Lq \rightarrow LqZ_i+A$), where Lq is the molecule of liquid solvent.

At high additive concentrations and at temperatures above T_g the simple law (1) is violated and the additive concentration in the polymer becomes greater than $[Z_a]$ calculated at moderate A concentrations [3,7,8]. To explain this phenomenon consider the energy of the sorption center formation.

The elements of order violation are in the same time the sites in which the polymer density is decreased, otherwise they represent the elements of free (excess) volume. To form an excess volume v_i in the polymeric matrix the certain energy is needed. We assumed that this energy is directly proportional to the volume formed, i.e. [9] that:

$$\varepsilon_i=qv_i \tag{6}$$

for all meanings of i. The dimension of coefficient q is energy/volume or pressure. In the presence of low-molecular compound A the value of q will decrease by the value of energy of interaction between the molecules of A surrounding polymer q_a, i.e. q must be substituted with $q-q_a$.

It follows from (6) that $\Delta\varepsilon_i/\Delta v_i=q-q_a$ does not depend on both v_i and Δv_i values. It means that if the center Z_i already exists and its volume is less than that necessary for sorption of A molecules v_a, to increase the value v_i to v_a, that is - to rearrange this center, the certain minimal (threshold) value of osmotic pressure of the additive $P_A=q-q_a$ is needed which does not depend on the size of v_i. In the polymer melt, when all volume of the polymer sample is accessible for mobile part of the additive, this osmotic pressure P_A is connected with A concentration around polymer $[A]_m$ by expression:

$$PA=RT[A]=RT\phi[A]_p \tag{7}$$

where $\phi=[A]/[A]_p$ - the variable coefficient which may be calculated from expression (5). When $[A]_p<<[Z_a]$ $\phi=(K_a[Z_a])^{-1}$, but deviates from this when $[A]_p$ is comparable with $[Z]_a$. It follows from (7) that the minimal additive concentration is needed above which the sorption centers will rearrange. If $q_a \geq q$ so $P_A \leq 0$ and the polymer will be swelling at any concentration of A.

Transformation of the element of disorder ζ_i into new sorption center Z^*_{ai} may be written as:

$$\zeta_i + A \rightarrow Z^*_{ai}A \tag{8}$$

Assuming that because of low translation mobility in solid polymer these new centers after removal of A must exist as relatively stable structures, we studied the effect of high concentrations of phenyl-β-naphthylamine, $C_{10}H_7$-NH-C_6H_5, and phenyl benzoate, C_6H_5-CO-O-C_6H_5, dissolved in isotactic polypropylene melt, on solubility of the same and other additives in the solid polymer prepared from this melt. Considering the additive which changes the structure of polymer as plasticizer, we may name the changes of the polymer structure remaining after removal of this additive post-plasticizing.

EXPERIMENTAL

Isotactic polypropylene ProFax 6501 produced by Himont Co (U.S.A.), molecular mass M_w=164 400, M_w/M_n=38, washed with heptane from atactic fraction, and additives: phenyl benzoate (PB), molecular mass M=198.2 and phenyl-β-naphthylamine (PNA), M=219.3, crystallized from heptane, were used. The samples were prepared by heating the mixtures of polypropylene with phenyl benzoate to 220-230° between two polished plates of stainless steel with subsequent pressing into films 0.005 cm thick and rapid cooling to room temperature. For analysis of additives concentration the latter were extracted with alcohol (48 hours, 40°C), the additive quantity being determined spectrophotometrically [10].

Experiments on additives solubility were performed at 40°C.

RESULTS AND DISCUSSION

Variation of concentrations of PNA and PB in the polypropylene samples with various initial concentrations of these compounds kept in the saturated vapors of the same compound at 40° is shown in Figures 1 and 2. As seen from these figures, the concentrations of both compounds in untreated polypropylene increases reaching the limits equal to 0.008 mol/kg for PNA and 0.055 mol/kg for PB, but the concentrations of both compounds in the samples prepared from polypropylene melt containing various concentrations of PNA and PB, decreased approaching the limits depending on additive concentration in the polymer melt.

Variation of additive concentration in the sample is the result of the additive diffusion from the sample surface inside or, vice versa, from inside the sample to its surface. The result of diffusion in uniform media is equalization of diffusant concentration in it, that is when the average additive concentration in the sample becomes constant it means that its concentration through the ample is the same and is in equilibrium with the additive vapors around the sample.

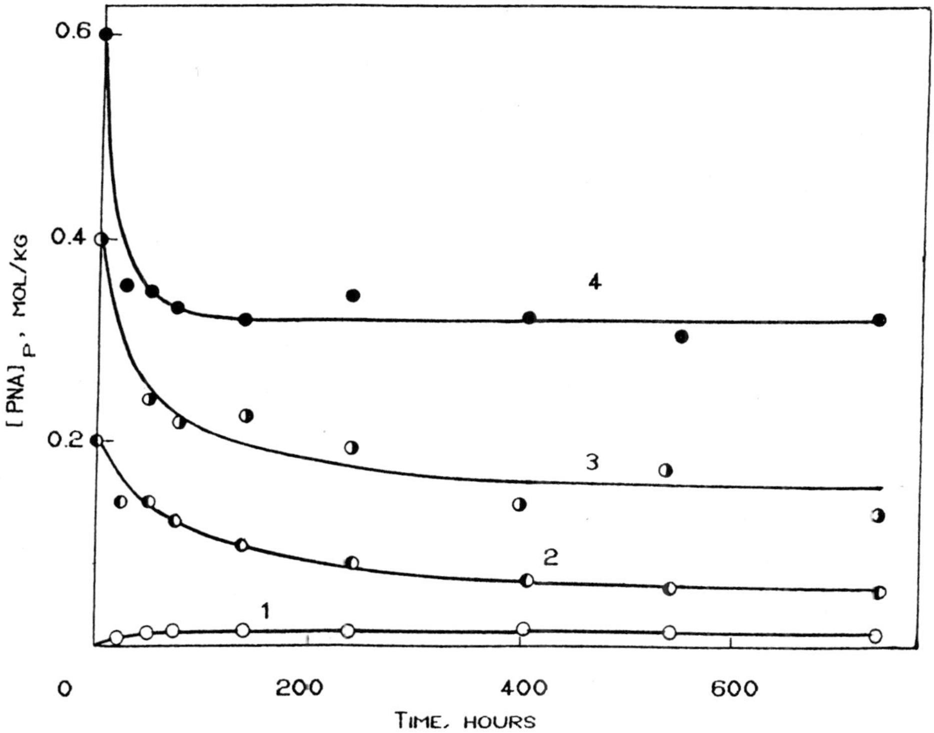

Figure 1. Variation of concentration of PNA $[PNA]_p$ in non-treated polypropylene (1) and in polypropylene samples, prepared from melts containing PNA, 0.2 92), 0.4 (3) and 0.6 mol/kg (4). 40°C.

As seen from Figure 3, to increase the additive solubility the concentration of this additive in the polymer melt must exceed the certain limiting value. This is in the good agreement with the theory considered above according to which the certain minimal additive concentration is needed to rearrange the sorption center. Now we have enough data for prediction of these limiting concentration values.

The increase in additive solubility may be due to increase of sorption centers concentration $[Z_a]$, or to increase in the meaning of equilibrium constant K_a (or of the product $\gamma_a K_a$). To distinguish between these possibilities we studied the sorption isotherms of both PNA and PB by non-treated and treated polypropylene from heptane solutions of the additives. The results of these experiments are presented in Figures 4 and 5.

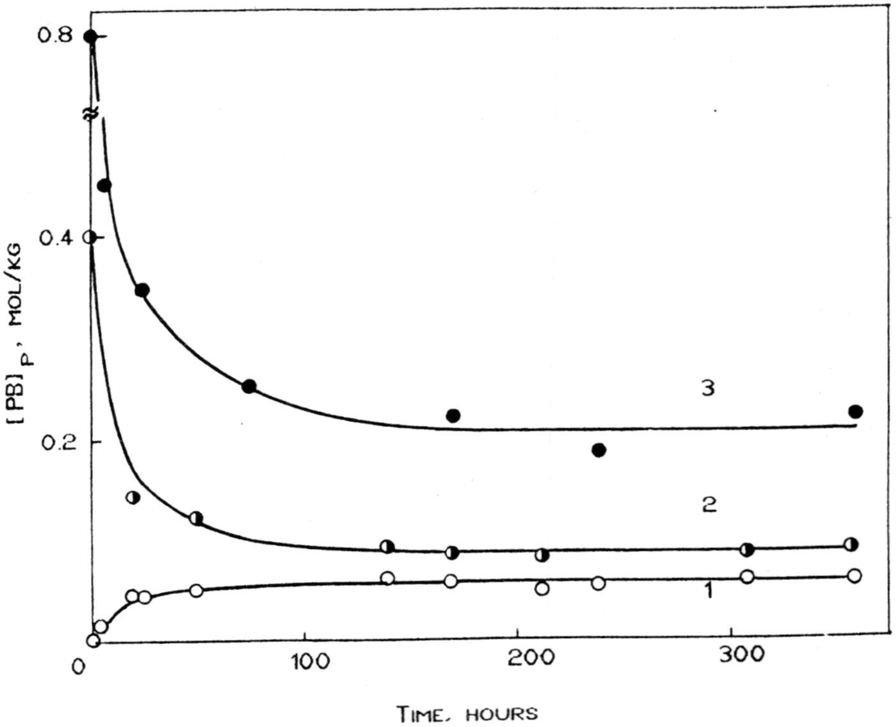

Figure 2. Variation of concentration of PB $[pb]_p$ in non-treated polypropylene (1) and in polypropylene samples, prepared from melts containing PB, 0.4 (2) and 0.8 mol/kg (3). 40°C.

All the isotherms presented in Figures 4 and 5 can be described by formula (5). The values of $[Z_a]$ and $\gamma_a K_a$ calculated from data presented in these figures were: for A=PB $[Z_a]=1.3 \cdot 10^{-2}$ mol/kg, $\gamma_a K_a=64$ l/mol in non-treated polypropylene and $[Z_a]=1.8 \cdot 10^{-2}$ mol/kg, $\gamma_a K_a=47$ l/mol in polypropylene treated in the molten state with PB, 0.8 mol/kg; for A=PNA $[Z_a]=2.3 \cdot 10^{-2}$ mol/kg, $\gamma_a K_a=101$ l/mol in non-treated and $[Z_a]=12.0 \cdot 10^{-2}$ mol/kg, $\gamma_a K_a=32$ l/mol in polypropylene treated with PNA, 0.6 mol/kg. It must be noted that increase in sorption centers concentration is accompanied with decrease in the values of constants of sorption equilibrium: the sorption centers Z_a^* formed in the polymer melt according to process (7) are less "convenient" then the pre-existing in the polymer. Thus not changing the number of UODs but changing the structures of some of them we can change the level of disorder in the polymer [11].

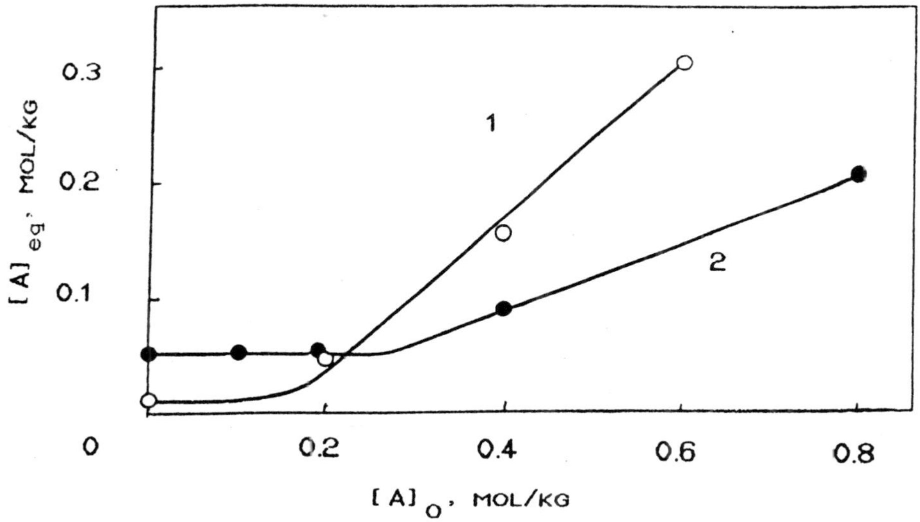

Figure 3. Equilibrium additive concentrations in isotactic polypropylene at 40°C $[A]_p$ as function of its concentration in the polymer melt $[A]_0$.A=PNA (1) and A=PB (2).

Table 1. Solubilities of various compounds in isotactic polypropylene, non-treated and treated with phenyl benzoate in the melt (mol/kg). 40°C.

	Compound			
	Phenyl benzoate	**diphenylamine**	**dibenzylsulfide**	**phenyl-β-naphthylamine**
Non-treated	0.055	0.11	0.35	0.008
Treated with PB, 0.6 mol/kg	0.15	0.27	0.83	0.032

As seen from Table 1, the centers prepared by treatment of polymer with one additive can sorb some other ones: the solubilities of diphenylamine, dibenzylsulfide and PNA in isotactic polypropylene prepared from melt containing PB were markedly greater than in non-treated polymer.

Opposite to the centers existing in non-treated polymer, the new-formed centers Z_a^* are unstable and disappear at the moderate heating of the sample. Figures 6 and 7 show the decrease of solubilities of PB and PNA, measured at 40°C, as functions of heating time at 80° and 100°C. Before heating all samples were washed from additives. As seen from these figures, the heating at 80°C decreases the solubilities of both additives to the values

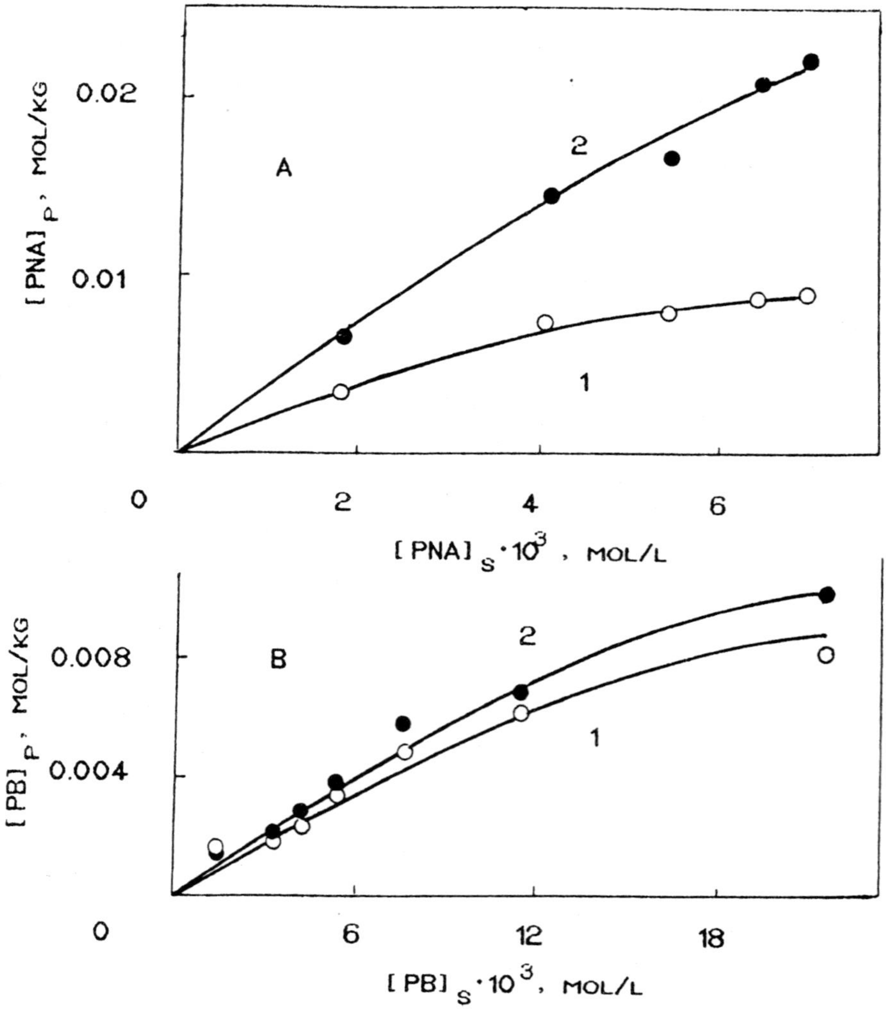

Figure 4. Sorption isotherms of PNA (A) and PB (B) from heptane solutions by non-treated polypropylene (1) and polypropylene samples prepared from melts, containing PNA, 0.6 mol/kg, and PB, 0.8 mol/kg. 40°C, $[PNA]_S$ and $[PB]_S$ - the concentrations of additive in heptane solution.

markedly differing from their solubilities in non-treated polymer, when the heating at 100° results in nearly complete disappearance of new-formed centers due to relaxation of the substance surrounding UOD. The process can be formally described by equation:

$$Z_a^* \rightarrow \zeta_i \qquad (9)$$

It follows from (5) that the concentration of mobile molecules of the additive A at $[A]_p << [Z_a]$ is $[A]=[A]_p/K_a[Z_a]$ or, if the polymer sample was prepared from melt containing high A concentration, $[A]=[A]_p/K^*_a[Z^*_a]$: the mobility of the additive and, if A is antioxidant, the apparent rate constant of its interaction with virtually immobile macroradicals which, besides other factors, is proportional to the ratio $\phi=[A]/[A]_p$, can depend on the way of the additive mixing with the polymer.

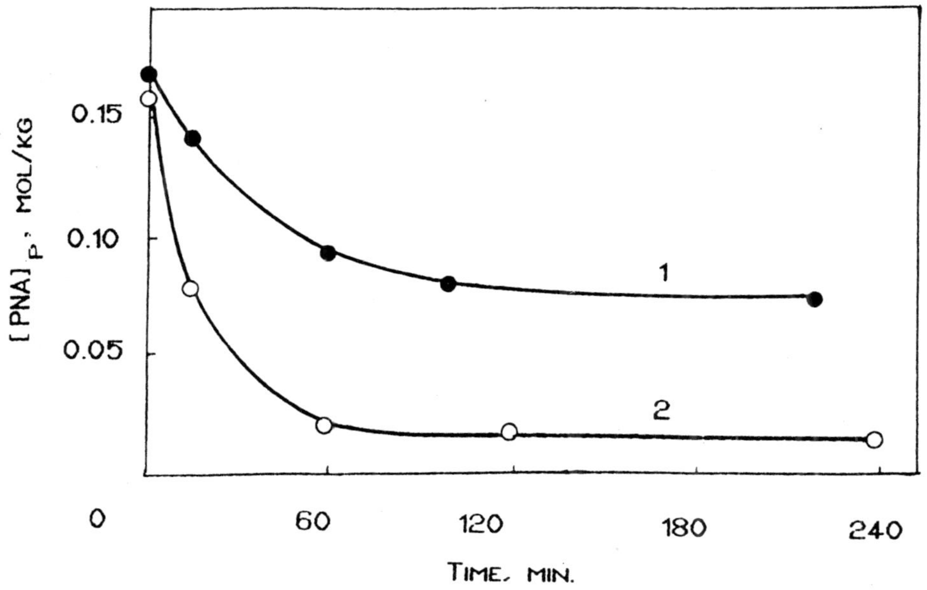

Figure 5. Variation in PNA solubility $[PNA]_p$ in polypropylene samples prepared from melt containing PNA, 0.6 mol/kg, as function of time of heating at 80° (1) and 100°C (2).

As shown in this paper, some elements of polymer structure may be reversibly changed without changing UOD topology and concentration. When these new-formed centers contain molecules of an additive A polymer may be considered as plasticized by A, but the changes preserved after removal of this additive may be considered as post-plasticizing.

As shown in [12], the concentration and composition of UODs in a polymer and, as a sequence, properties of the polymer can be varied by variation of condition of the polymer sedimentation. The difference between samples prepared from different solvents remains even in the polymer melt.

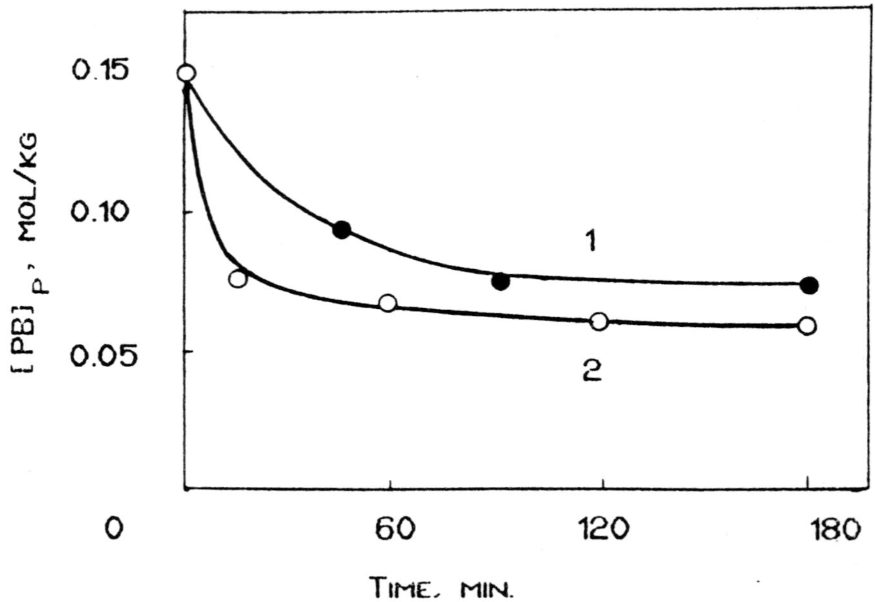

Figure 6. Variation of PB solubility $[PB]_p$ in polypropylene samples prepared from melt containing PB, 0.8 mol/kg, as function of time of heating at 80° (1) and 100°C (2).

REFERENCES

1. Shlyapnikov Yu.A. Dokl Akad. Nauk SSSR, 202, 1377 (1972).
2. Shlyapnikov Yu.A. Europhys. Conf. Abstr. 51, 311 (1981).
3. Shlyapnikov Yu.A. Sov. J. of Chem. Phys. 11, 1898 (1991).
4. R.M. Barrer, J.M. Barrie, J. Polym. Sci. 23, 331 (1957).
5. Shlyapnikov Yu.A., Mar'in A.P. Eur. Polym. J. 23, 623 (1987).
6. Sataite J., Mar'in A.P., Juskeviciute S., Shlyapnikov Yu.A., Vysokomolek. soedin., B, 27, 215 (1985).
7. Bogaevskaya T.A., Shlyapnikov Yu.A., Dokl. Akad. Nauk SSSR, 210, 1362 (1972).
8. Mar'in, A.P. Shlyapnikov Yu.A., Doklady Akad. Nauk SSSR, 219, 1409 (1974).
9. Shlyapnikov Yu.A., Polymer, 35, 1031 (1992).
10. Juskeviciute S., Shlyapnikov Yu.A., Lietuvos TSR Mokslu akademijos darbai (Proceedings of Lietuvian Academy of Sciences), B, 3 (58), 153 (1969).
11. Shlyapnikov Yu.A., Intern. J. Polymeric Mater., 19, 101, (1993).
12. Shlyapnikov Yu.A., Monakhova T.V., Bogaevskaya T.A., Polym. Degradation & Stability, 46, 247 (1994).

Theoretical Estimation of Acid Strength of Boron-Fluoride-Alcohol Complexes During Cationic Polymerization of Olefins

*K.C. Minsker, V.A. Babkin and G.E. Zaikov**
Bashkirian State University, Ufa-750071, Frunze st. 32
**Institute of Chemical Physics Russian Academy of Sciences*
Moscow, 117334, Kosigina st. 4.

Complexes of alcohols (CH_3CH, C_2H_5-CH, Izo-C_3H_7, t-C_4H_9) and boron fluoride BRnF3-n (here n=0, 1, 2, 3; R=CH_3, C_2H_5, izo-C_3H_7, t-C_4H_9) are the traditional acidic catalysts of the polymerization of olefins [1-3]. Changes of the nature of proton donors-alcohols and of ligand environment of B provides the formation of initiating systems with a wide range of activity.

The labile character of these systems proved by changing of their nature and composition during storage or in the reacting medium and makes it difficult to carry out the analysis of catalytic and acidic properties. One of the important features which determines catalytic properties of ROH, BRnF3-n s the acidic strength on which their activity depends in various processes. As there are no reliable methods for experimentally determining the acidic strength (pKa) of H-complexes [2], an important role is played by the theoretical methods of estimating such characteristics of these systems as the charge on H-atom ROH-q_H+, which correlates with the universal exponent of acidity pKa [4].

Dependence of the electrophilic activity of alcohols in complexes on the acidic strength and the resulting possibility of demonstrating both protonodonor and carbocationic activities under the influence of Lewis and Brönsted acids allow establishing on the basis of theoretical quantum-chemical calculations, the correlation between acidic and catalytic properties of boron fluoride-alcohol complexes.

The quantum-chemical calculations of ROH. BRnF3-n-complexes have not been done, their acidic strength was not estimated, and its influence on the catalytic activity was not studied.

The quantum-chemical investigation of ROH.BRnF3-n-complexes is done in this work and their acidic strength is evaluated in dependence of the nature of alcohols and ligand environment of heteroatom B; the data is analyzed and systematized.

Models and Procedure of Calculations

Quantum-chemical calculations of ROH.BRnF3-n-complexes were done by CNDO/2 method in classic parametrization of Santry-Pople-Segal [5,6]. The optimization of geometry of the complexes under study was done by changing metrics [7]. The structure of complexes for all models was taken as tetrahedronal one, according to minimum energy of the main system state. The values of initial lengths of connections and angles between them were taken from [7].

The calculation was done in approximation of the isolated molecule in the gas phase. The parameters of acidic strength of complexes were represented by the charges of H^+ atoms of ROH-alcohols and the values of the universal exponent of acidity pKa. Bond order for the alcohols in the complexes with boron fluoride P_{OH} and P_{OC} in the estimation of the influence of acidic strength on manifestation by alcohols of proton-donor / carbocationic activity were determined according to Armstrong [8].

Results of Calculations and Their Discussion

Table 1 gives the parameters of acidic strength (q_{H+} and pKa), bond orders P_{OH} and P_{OC}, and corresponding activities -H^+ (protodonor) or R^+ (carbocationic) of alcohols in ROH. BRnF3-n systems.

If both H^+ and R^+ stand in one column of the table, then it means that carbocationic and protodonor activities of alcohols are equally possible. For easy comparison of results of calculations all 52 models under study are subdivided into 13 groups which differ from each other by the nature of ligand environment of boron. Electronic structure and geometry of boron fluoride-alcohol models after geometry optimization by alternative metric method represent distorted tetrahedronal configurations similar by their characteristics to those represented in work [7]. (Bond length of R_{OH}=0.1 nm, R_{BF}=0.135 nm, R_{OB}=(0.17±0.005)nm, R_{OH}=0.109 nm, R_{CC}=(0.132±0.004)nm. Angles between ligands B=109±18°, valent to angles of alcohol HOR=114±10°). All the models ROH.BRn F3-n represent donor-acceptor complexes. Quantum-chemical proof of the fact is high negative values of the total energy of the systems

(Ec=302590-370990 kj/mol.) characterizing the stability of complexes. The part of the charge of electrone transmitted to the receptor was equal to 0.07-0.09.

Table 1. Parameters of acidic strength and bond orders P_{OH} and P_{OC} of the complexes $R_nBF_{3-n}{\cdot}ROH$.

N n/n	Model	q_{H+}(±0.02)	pKa	P_{OH}	P_{OC}	Activity	N group
1	$F_3B.OHCH_3$	+0.24	+1.0	0.92	0.98	H^+	
2	$F_3B.OHC_2H_5$	+0.23	+1.5	0..92	0.97	H^+	I
3	$F_3B.OHC_3H_7$	+0.21	+7.1	0.92	0.94	H^+, R^+	
4	$F_3B.OHC_4H_9$	-0.20	+7.3	0.93	0.90	R^+	
5	$F_2CH_3B.OHCH_3$	+0.23	+1.5	0.92	1.01	H^+	
6	$F_2CH_3B.OHC_2H_5$	+0.22	+4.4	0.92	0.96	H^+	II
7	$F_2CH_3B.OHC_3H_7$	+0.20	+7.3	0.93	0.90	R^+	
8	$F_2CH_3B.OHC_4H_3$	+0.19	+7.7	0.93	0.89	R^+	
9	$F(CH_3)_2B.OHCH_3$	+0.23	+1.5	0.92	1.00	H^+	
10	$F(CH_3)_2B.OHC_2H_5$	+0.21	+7.1	0.94	0.97	H^+	III
11	$F(CH_3)_2B.OHC_3H_7$	+0.20	+7.3	0.92	0.94	H^+, R^+	
12	$F(CH_3)_2B.OHC_4H_3$	+0.20	+7.3	0.93	0.90	R^+	
13	$(CH_3)_3B.OHCH_3$	+0.24	+1.0	0.94	0.96	H^+, R^+	
14	$(CH_3)_3B.OHC_2H_5$	+0.23	+1.5	0.92	0.95	H^+	IV
15	$(CH_3)_3B.OHC_3H_7$	+0.22	+4.4	0.94	0.91	R^+	
16	$(CH_3)_3B.OHC_4H_9$	+0.20	+7.3	0.92	0.89	R^+	
17	$F_2C_2H_5B.OHCH_3$	+0.23	+1.5	0.93	0.98	H^+	V
18	$F_2C_2H_5B.OHC_2H_5$	+0.22	+4.4	0.92	0.96	H^+	
19	$F_2C_2H_5B.OHC_3H_7$	-0.22	+4.4	0.94	0.93	H^+, R^+	
20	$F_2C_2H_5B.OHC_4H_9$	+0.21	+7.1	0.93	0.89	R^+	
21	$F(C_2H_5)_2B.OHCH_3$	+0.24	+1.0	0.92	0.97	H^+	VI
22	$F(C_2H_5)_2B.OHC_2H_5$	+0.23	+1.5	0.92	0.95	H^+	
23	$F(C_2H_5)_2B.OHC_3H_7$	+0.21	+7.1	0.92	0.93	H^+, R^+	
24	$F(C_2H_5)_2B.OHC_4H_9$	+0.20	+7.3	0.93	0.90	R^+	
25	$(C_2H_5)_3B.OHCH_3$	+0.22	+4.4	0.92	0.90	H^+	
26	$(C_2H_5)_3B.OHC_2H_5$	+0.21	+7.1	0.93	0.95	H^+, R^+	VII
27	$(C_2H_5)_3B.OHC_3H_7$	+0.21	+7.1	0.92	0.93	H^+, R^+	
28	$(C_2H_5)_3B.OHC_4H_9$	+0.20	+7.3	0.93	0.90	R^+	

Table 1 (Continued)

N n/n	Model	q_{H+}(±0.02)	pKa	P_{OH}	P_{OC}	Activity	N group
29	$F_2C_3H_7B.OHCH_3$	+0.23	+1.5	0.92	0.98	H^+	
30	$F_2C_3H_7B.OHC_2H_5$	+0.22	+4,4	0.93	0.97	H^+	VIII
31	$F_2C_3H_7B.OHC_3H_7$	+0.20	+7.3	0.92	0.93	H^+, R^+	
32	$F_2C_3H_7B.OHC_4H_9$	+0.20	+7.3	0.92	0.89	R^+	
33	$F(C_3H_7)_2B.OHCH_3$	+0.24	+1.0	0.92	0.99	H^+	
34	$F(C_3H_7)_2B.OHC_2H_5$	+0.24	+1.0	0..93	0.96	H^+	IX
35	$F(C_3H_7)_2B.OHC_3H_7$	+0.21	+7.;1	0.93	0.94	H^+, R^+	
36	$F(C_3H_7)_2B.OHC_4H_9$	+0.20	+7.3	0.92	0.89	R^+	
37	$(C_3H_7)_3B.OHCH_3$	+0.22	+4.4	0.92	0.99	H^+	
38	$(C_3H_7)_3B.OHC_2H_5$	+0.22	+4.4	0.92	0.96	H^+	X
39	$(C_3H_7)_3B.OHC_3H_7$	+0.21	+7.1	0.92	0.93	H^+, R^+	
40	$(C_3H_7)_3B.OHC_4H_9$	+0.19	+7.7	0.92	0.89	R^+	
41	$F_2(C_4H_9)B.OHCH_3$	+0.23	+1.5	0.92	0.99	H^+	
42	$F_2C_4H_9B.OHC_2H_5$	+0.21	+7.1	0.93	0.96	H^+	XI
43	$F_2C_4H_9B.OHC_3H_7$	+0.21	+7.1	0.92	0.93	H^+, R^+	
44	$F_2C_4H_9B.OHC_4H_9$	+0.20	+7.3	0.93	0.89	R^+	
45	$F(C_4H_9)_2B.OHCH_3$	+0.23	+1.5	0.92	1.00	H^+	
46	$F(C_4H_9)_2B.OHC_2H_5$	+0.22	+4.4	0.93	0.97	H^+	XII
47	$F(C_4H_9)_2B.OHC_3H_7$	+0.22	+4.4	0.93	0.94	H^+, R^+	
48	$F(C_4H_9)_2B.OHC_4H_9$	+0.19	+7.7	0.93	0.90	R^+	
49	$(C_4H_9)_3B.OHCH_3$	+0.22	+4.4	0.93	1.00	H^+	
50	$(C_4H_9)_3B.OHC_2H_5$	+0.21	+7.1	0.93	0.96	H^+	XIII
51	$(C_4H_9)_3B.OHC_3H_7$	+0.20	+7.3	0.93	0.93	H^+, R^+	
52	$(C_4H_9)_3B.OHC_4H_9$	+0.19	+7.7	0.93	0.91	H^+, R^+	

Transition of unbound electrones of oxygen ROH is performed by cRc-orbits to vacant to Sp^2 orbital B^{3+}.

One can see from the table that the charge of hydrogen atom-q_{H+} of alcohols in complexes with BRnF3-n has the tendency to reduce from 0.24 to 0.19 at the branching of alcohol radicals in all 13 groups. It means that the charge q_{H+} does not depend on the nature of ligand environment of B and is determined by the nature of the alcohol radical R.

The acidic strength of complexes in accordance with correlative relation q_{H+}=pKa [4] reduces from 0.5-1.5 to 7.3-7.7 in the ow of alcohols from CH_3OH to C_4H_9OH in fluoride boron complexes.

Bond order for all studied groups of complexes ROH.BRnF3-n is practically unchanged and equals 0.94-0.92 (possible mistake of CNDO/2 method at the estimation of order of connections is approximately 0.02). The order of bonds P_{OC} has tendency to reduction simultaneous with reduction of the acidic strength. The comparison of bond orders of P_{OH} and P_{OC} shows that for CH_3OH and C_2H_5OH in complexes with boron fluoride the protonodonor activity is characteristic, and R^+ is characteristic of tert-butyl one. In izo-propile alcohol both H^+ and R^+ activities are equally probable.

Two alternatives of the alcohol molecule fragmentation depending on the nature of the radical are experimentally proved for complexes ROH.BF3 [3]. Quantum-chemical calculation testifies to the fact that the analogical possibilities of ROH fragmentation are characteristic of complexes with boron-organic compounds BRnF3-n. Moreover, for models ROH.BRnF3-n, where the nature of radical ligand environment of B coincides with the nature of alcohol radicals, i.e. for complexes $(CH_3)3B.\ OHCH_3$, $C_2H_5)\ 3B.OHC_2H_5$, $(C_3H_7)\ 3B.OHC_3H_7$, $(C_4H_9)3B.OCH_4H_9$ the equal probability of developing H^+ and R^+ activities is characteristic.

One of the reasons explaining this result is the coincidence of groups of symmetry of the alcohol radical and of heteroatom B, leading to equalization of bond orders P_{OH} and P_{OC}. However, this fact requires experimental check.

Thus, according to the data of quantum-chemical calculations the ROH.BRnF3-n complexes have low Brönsted acidity (+0.5 pKa +7.7) and do not depend on the nature of ligand environment B. To some degree, the demonstration by alcohols in complexes of protodonor and carbocationic activities in the reaction of polymerization is explained by the nature of alcohol radical. In the final analysis the low acidic strength of complexes supposes the relation between acidic-catalytic properties of ROH.BRnF3-n and the interaction with substrate.

REFERENCES

1. J. Kennedy, Cationic Polymerization of Olefins: A Critical Inventory, New York, 430 pp. (1975).
2. Alah, C.A. Friedel-Crafts Chemistry, New York, 581 pp. (1973).
3. Dainton, F.C., Sutherland, C.B.B.M. J. Polon. Sci., 1949, 4, Ni, pp. 37-43.
4. Grabowski, W., Misono, M., Yoneda, Y. 1980, 61, N1, p. 103-108.
5. Pople, J.A., Beveridge, D.Z. Approximate molecule original theory. N.Y.: Mc Graw-Hill Book, 1970, p. 44.
6. Santry, P.P., Segal, G.A. J. Chem., Phys., 1967, 47, N1, p. 158-174.
7. Ezner, U.E., Erusalimski, B.L., Electronical aspect of the reaction polymerization. L.: Science, 1976, 180 p.
8. Armstrong, D.R., Perkins, P.S., Stewart, J.J.P., J. Chem. Soc. Dolton Trans., 1973, 8, p. 838-840.

Synthesis of Epoxicontaining Siliconorganic Compounds

Lotari M. Khananashvili, David A. Girgvliani, Nodari I. Tsomaya, and Tsisana N. Vardosanidze

Tbilisi State University, 3 Chavchavadze Ave. 380028 Tbilisi, Republic of Georgia

Abstract—Epoxicompounds containing the methyl, phenyl, vinyl, chlormethyl, thienyl and chlorthienyl radicals with molecules containig oen siolxane or cyclosiloxane, or two and more siloxane fragments are synthesized for the first time by the hydrosililization of alylglicidil ester with organohydrosilanes or siloxanes in the presence of Speier catalizator H_2PtCl_6. The reaction of hydrosililization is studied using infrared spectroscopy and gas-liquid chromatography methods. The anionic polymerization of epoxicompounds containing cyclosiloxane fragments using α,β-bis (tetramethilamonium) dimethilsilanolat is carried out.

Experimental Section

1,3-dichlortetramethyldisiloxane, Boiling temperature 138°C, %Cl-34.60, n^{20}_D=1.4050 (1), 1,5-dichlorhexamethyltrisiloxane, Boiling temperature 80°C/20 mm mercury column %Cl-25.60, n^{20}_D=1.4070, 1,3-dichloro-1,3-dimethyl-1,3-diphenyldisiloxane, Boiling temperature 154-155°C/1 mm mercury column %Cl-21.64, n^{20}_D=1.5719, 1,3-dichloro-1,3,5-trimethyl-1,3,5-triphenyltrisiloxane, Boiling temperature 189°C/1mm mercury column %Cl-15.40, n^{20}_D=1.5370, 1,3-dichloro-1,3-dimethyldisiloxane, Boiling temperature 103°C %Cl-40.48, %H(Si-H) 1.15, were obtained by the partial hydrolysis of corresponding organolorsilanes using the accepted methodics.

The synthesis of heptamethylcyclotetrasiloxane. 120ml of absolute ester and 31.6g (0.4 mole) of pyridine was poured into a four-neck flask equipped with two dropping funnels, reverse refrigerator and mechanical stirrer. 48g (0.2 mole) of dihydroxihexamethiltrisiloxane was added from one dropping funnel, while 23 g (0.2 mole) of methildichlorsilane from other dropping funnel, dissolved in 120 ml of absolute ester.

The whole reaction mixture was stirred and when temperature achieved 30°C, was heated during the 5 hours. After that the solution was cooled and sediments were filtered. The ester solution was washed off by the distilled water till the neutral reaction and was dried on waterless sodium sulphur. Volatile products were distilled under vacuum. The rest of reaction products were distilled. The boiling temperature is 165°C %H(Si-H)-0.36, n^{20}_D=1.3965. This corresponds to ref. (2) 2,4,6-trimethyl-2,4-diphenylcyclotrisiloxane was obtained analogically. The boiling temperature is 160-162° C/1mm mercury column, %H(Si-H)-0.28. This corresponds to ref. (2) 2,4,6,8-tetramethyl-2,4,6-triphenylcyclotetrasiloxane was obtained analogically. The boiling temperature is 190-191°C/1mm mercury collumn, %H(Si-H)-0.20.n^{20}_D=1.5330. This corresponds to ref. (2) 2,2,4,4,6,8-hexamethylcycllotetrasiloxane was obtained analogically. The boiling temperature is 139-153°C/1mm mercury column, %H(Si-H) - 0.51. n^{20}_D=1.4905.

The synthesis of glisidoxipropilheptamethylcyclotetrasiloxane. 27g (0.1 mole) of heptamethylcyclotetrasiloxane and 0.05ml 0.1N solution of platinumclorhydrogen acid in propanol was poured into a three-neck flask equipped with dropping funnel, thermometer and mechanical stirrer. 12.5g (0.11 mole) of aliglglicidil ester was added from dropping funnel. After the growth dropping funnel was replaced by reverse refrigerator and the reaction mixture was heated to 70-80°C during 3 hours. Then the volatile products were distilled. Finally by the distillation glisidoxipropilheptamethyl-cycllotetrasiloxane was obtained. The boiling temperature is 268-270°C/2mm mercury collumn n^{20}_D=1.4432, the content of epoxi-groups is 10.56% (calculated - 10.85). 6-(glicidoxipropil)-2,4,6-trimethyl-2,4-diphenylcyclotrisiloxane was obtained analogically. The boiling temperature is 240-245°C/1mm mercury collumn n^{20}_D=1.4453, the content of epoxi-groups is 9.95% (calculated - 9.64). 8-(glicidoxipropil)-2,4,6,8-tetramethyl-2,4,6-triphenylcyclotetrasiloxane was obtained analogically. The boiling temperature is 100-104°C/10 mm mercury collumn n^{20}_D=1.4558, the content of epoxi-groups is 7.82% (calculated - 7.39). 6,8-bis(glicydoxipropil)-2,2,4,4,6,8-hexamethilcyclotetrasiloxane was obtained analogically. The boiling temperature is 127-129°C/10 mm mercury collumn n^{20}_D=1.4530, the content of epoxi-groups is 17.91% (calculated 17.34). 6,8-bis(glicydoxipropil)-2,4,6,8-tetramethyl-2,4-diphenylcyclotetrasiloxane was obtained analogically. The boiling temperature is 167-169°C/10 mm mercury collumn n^{20}_D=1.4472, the content of epoxi groups is 14.46% (calculated 13.87).

RESULTS AND DISCUSSION

In literature there are basicly known the siliconorganic epoxi-monomers with methyl and phenil radicals bounded to silicon atom, containing one or two silicon atoms in chain. The oligomeric epoxisiloxanes containing a small quantity of epoxigroups has also been synthesized.

Our group was the first who synthesized the epoxi-compounds containing siloxane fragments with methyl, vinyl, chlorethyl, thienyl and chlorthienyl radicals. These compounds are synthesized by the hydrosililization of aliglicidil ester with hydridsilanes or siloxanes in the presence of Speier catalizator (H_2PtCl_6) in inert area with 80°C. By the hydrosililization of alylglicidil ester with dimethylthienilsilane, dimethylchlorthienylsilane, thienyldichlorsilane and chlorthienyldichlorsilane the compounds containing one epoxi-groupd in molecule are obtained.

$$RR'_2SiH + CH_2 = CHCH_2OCH_2\underset{\setminus O /}{CH - CH_2} \xrightarrow{H_2PtCl_6} RR'_2Si9CH_2)_3OCH_2\underset{\setminus O /}{CHCH_2}$$

where

$$R = \sqrt{S}, R' = Me(I); R = Cl\sqrt{S}, R' = Me(II); R = \sqrt{S},$$
$$R = Cl(III); R = C\sqrt{S}, R' = Cl(IV); R = \sqrt{S}, R' = OC_2H_5(V);$$
$$R = Cl\sqrt{S}; R' = OC_2H_5(VI).$$

the characteristics of obtained compounds are given in Table 1.

To obtain the diepoxides containing one or more silicon atom, the following scheme was used: first by the partial hydrolysis of organodichlorsilanes we have obtained the α,ω-dichlorodiorganosiloxanes and then by its fractionization we have collected the organosiloxanes containing two or three silicon atoms, and by its regeneration with $LiAlH_4$ we were separating of alilglicidyl ester and dihydrosiloxanes can be described by:

$$H(SiRR'O)_n - \underset{R'}{\overset{R}{Si}} - H + 2CH_2 = CHCH_2)CH_2\underset{\setminus O /}{CHCH_2} \xrightarrow{H_2PtCl_6}$$

$$\longrightarrow \underset{\setminus O /}{CH_2CH} - CH_2O(CH_2)_3 - \left(\underset{R'}{\overset{R}{Si}} - O\right)_n - \underset{R'}{\overset{R}{Si}} - (CH_2)_3OCH_2\underset{\setminus O /}{CH \cdot CH_2}$$

Table 1. Physical and chemical properties and elementary content of synthesized epoxicyclosiloxanses

N	B.P. °C	n_D^{20}	d_{20}^{4}	MR		Elementary Obtained				FORMULA	Elementary Calculated			
				Obtained	Calculated	Si	S	Cl	$-CH-CH_2$ (epoxy, -O-)		Si	S	Cl	$-CH-CH_2$ (epoxy, -O-)
I	131-133 2 MM	1.4590	0.9337	73	73.51	10.20	12.17	-	16.20	$Si_1C_{12}H_{20}O_2S$	10.93	12.50	-	16.80
II	140-142 2 MM	1.4927	1.0671	78	78.35	10.14	11.43	11.36	15.08	$Si_1C_{12}H$[illegible]ClS	9.63	11.02	12.22	14.80
III	148-149 2 MM	1.4994	1.2036	72	72.77	9.61	11.08	24.56	14.33	$Si_1C_{10}H_{14}O_2Cl_2S$	9.42	10.77	24.14	14.48
IV	149-151 1.5 MM	1.4990	1.2428	78	77.61	8.64	10.24	31.86	12.32	$Si_1C_{10}H_{13}O_2Cl_3S$	8.44	9.65	32.13	12.97
V	155-156 2MM	1.5040	1.1205	83	84.29	8.26	11.12	-	13.93	$Si_1C_{14}H_{24}O_4S$	8.86	10.13	-	13.61
VI	170-172 2MM	1.4992	1.1642	88	88.20	8.24	10.02	9.25	12.86	$Si_1C_{14}H_{23}O_4ClS$	7.99	9.13	10.12	12.26
VII	182-183 1MM	1.4410	0.9942	115	116.68	19.64	-	-	20.08	$Si_3C_{18}H_{40}O_6$	19.27	-	-	19.72
VIII	178-180 1MM	1.4521	0.9844	106	106.36	14.97	-	-	23.25	$Si_2C_{18}H_{34}O_5$	14.50	-	-	22.28
IX	211-212 1MM	1.4576	0.9945	129	129.16	17.21	-	-	18.80	$Si_3C_{21}H_{40}O_6$	17.80	-	-	18.22
X	210-212 1MM	1.4510	0.9599	137	137.56	11.70	-	-	16.72	$Si_2C_{26}H_{38}O_5$	11.52	-	-	17.69
XI	240 1MM	1.4572	0.9681	175	175.96	13.90	-	-	13.26	$Si_3C_{33}H_{46}O_6$	13.50	-	-	13.83
XII	194-196 1MM	1.45[illegible]4	0.0111	135	135.38	12.31	12.2	-	17.00	$Si_3C_{22}H_{34}O_5S_2$	11.24	12.85	-	17.27
XIII	240-242 1MM	1.5010	1.0873	172	172.69	14.10	14.2	-	14.06	$Si_3C_{27}H_{40}O_6S_3$	13.12	15.00	-	13.44

where

$$n = 2, R = R' = Me(VII); n = 1, R = Me, R' = Vin(VIII);$$
$$n = 2, R = Me, R' = Vin(IX); n = 1, R = Me, R' = Ph(X);$$
$$n = 2, R = Me, R' = Ph(XI); n = 1, R = Me, R' = \sqrt{S}(XII);$$
$$n = 2, R = Me, R' = \sqrt{S}(XIII).$$

The synthesized silicon-containing epoxides are transparent liquids, which can be distilled at low pressures. Their physical and chemical properties are given in Table 1.

After the distillation of obtained liquids by the method of gas-liquid chromatography the individual properties of synthesized mono and diepoxides are determined. Analysis was held by the chromatographic collumn, with chromosorb-carrier W, with SE-30 mark siliconorganic cauchuk 50% solution, and helium as gas-carrier.

Apart from the abovementioned siliconorganic epoxides, for the first time, by our group, the mono and diepoxides containig cyclosiloxanic fragments with methyl and phenyl radicals in the main chain are synthesized. In order to obtain the abovementioned compounds we have synthesized the organocyloxanes, which contained one or two hydrogen atom with silicon atom. These organocyclosiloxanes were obtained using the reaction of heterofunctional condensation (3). To obtain the six-component cyclosiloxanes with one Si-H functional group, the reaction of partial hydrolysis of diorganodichlorsilanes was held. By partial hydrolysis the 1,3-dihydroxidorganosiloxanes were separated, and by their complete hydrolysis 1,3-dihydroxidiorganosiloxanes were obtained. By the heterofunctional condensation of 1,3-dihydroxidorganosiloxanes with methyldichlorsilanes in the presence of pyridine, the six-component organocyclosiloxanes with one Si-H group was obtained.

Using the same scheme, we have obtained eight-component compounds with one Si-H group, 1,5-dihidroxidiorganosiloxanes were obtained by the partial hydrolysis, then by their complete hydrolysis the 1,5-dihidroxidorganosiloxanes were separated. By the heterofunctional condensation of the obtained compounds with methildichlorsilanes in the presence of pyridine, the eight-component organocyclosiloxanes with one Si-H group were obtained. The reaction can be described by the following formula:

$$HMeSiCl_2 + HO\left(\underset{R'}{\overset{R}{Si}}-O\right)_n H \xrightarrow[-2Py-HCl]{+2Py} Me-\overset{H}{Si}-(OSiRR')_n$$

where

$$n = 2.\ R = R' = Me; n = 2.\ R = Me.\ R' = Ph;$$
$$n = 3, R = R' = Me; n = 3, R = Me, R' = Ph.$$

The synthesis of cyclotetrasiloxanes with two Si-H groups. By the partial hydrolysis of methildiclorsilanes we have obtained 1,3-dichlordimethyldisiloxane. By its heterofunctional condensation with 1,3-dihidroxitetraorganosiloxanes we have obtained eight-component cycles with two Si-H groups:

$$\left(Cl-\overset{H}{\underset{Me}{Si}}-\right)O+\left(HO-\overset{R}{\underset{R'}{Si}}-\right)_2 O \xrightarrow[-2Py.HCl]{+2Py} \left(H-\overset{Me}{Si}-O\right)_2-(SiRR')_2O$$

where

$$R = R' = Me;\ R = Me,\ R' = Ph.$$

By the reaction of alilglicidil ester with organosiloxanes containing one Si-H group in the presence of H_2PtCl_6 at 80°C, were obtained the monoepoxides containing cyclosiloxane fragments described by general formula:

$$(RR'SiO)_n-\overset{Me}{Si}-(CH_2)_3OCH_2CH-CH_2$$

where

$$n = 3,\ R = R' = Me(XIV);\ n = 2,\ R = Me,\ R' = Ph(XV);$$
$$n = 3,\ R = Me,\ R' = Ph(XVI).$$

By the reaction of organocyclosiloxanes with two Si-H group with alilglicidil ester we have obtained the following diepoxides:

$$\left[CH_2CHCH_2O(CH_2)_3-\overset{Me}{Si}-O\right]_2-[-SiRR']_2O$$

where

$$R = R' = Me(XVII);\ R-Me,\ R' = Ph(XVIII)$$

The physical and chemical properties of synthesized epoxides are given in Table 2.

No	B.P. °C	n^{20}_D	d^{20}_4	Obtained		Calculated		Formula
				MR	$CHCH_2$	MR	$CHCH_2$	
XIV	268-270 2 MM	1.4431	1.0343	100.8	11.05	101.79	10.85	$Si_4C1_3H_{32}O_6$
XV	240-245 1 MM	1.4553	1.0097	120.0	9.95	119.88	9.64	$Si_3C_{21}H_{30}O_5$
XVI	100-104 10-2 MM	1.4558	1.1208	158.2	7.82	158.99	7.39	$Si_4C_{28}H_{38}O_5$
XVII	127-129 10-2 MM	1.4530	1.0809	123.0	17.91	123.68	17.30	$Si_4C_{18}H_{40}O_8$
XVIII	167-169 10-2 MM	1.4472	1.0102	163.2	14.46	163.98	13.87	$Si_4C_{28}H_{44}O_8$

The hydrosililization of alilglicidil ester with hexamethyltrisiloxane was studied by the method of infra-red spectroscopy. The decrease of intensity of lines of absorbtion 2125-2130 cm^{-1}, corresponding to Si-H group and 1600-1680 cm^{-1}, corresponding to C=C group, points to the fact, that the reaction of hydrosililization basically takes place with the participation of the double link. This is also proved by the fact that there was no 3400-3600 cm^{-1} line of absorbtion, corresponding to associated and nonassociated secondary hydroxile groups, lines of absorbtion 850, 917 cm^{-1} corresponds to epoxi-groups.

The reaction of hydrosililization was also studied with the method of gas chromatography. The analysis was held on LXM-8 with the following properties: Active ionization detector: 3m length and 4mm diameter metalic chromatographic column. Following chromatographic phases were used: SE-30, REOPLEX-400, and OV-17. The best fractionization was achieved with OV-17. The different temperature regimes were selected for different monomers. For example the best ractionization temperature for tetramethildisiloxane was 100°C, while for diphenilsilane this temperature was 130°C. The temperature of evaporator and detector was 250°C. In the taken samples the quantites of compounds not interacting in reaction was measured, and correspondingly the rate of transformation of reagents. The anionic polymerization of epoxiorganocyclosiloxanes with α,ω-bis (tetramethylamonium) dimethylsilanolat was performed, with amonium content 6.09% and silicon chain n=38. The optimal quantity of initiator was $5*10^{-2}$ mole/litre. Reaction was held on 80°C.

During the process of polymerization viscosity and the quantity of epoxigroups was determined. During first 15-20 minutes the increase of viscosity was observed (η=0.15), then the formation of space polymers takes place.

RESULTS

By the hydrosililization of alilglizidil ester with organohydridsiloxanes and hydridcyclosiloxanes in the presence of Speier catalizator H_2PtCl_6 the epoxi-compounds containing siloxane and cyclosiloxane fragments are separated. The reactions of hydrosililization are studied using the methods of infra-red and gas chromatography.

REFERENCE

1. K.A. Andrianov. Siliconorganic Compounds. Moscow, GHI, 1955, pp. 317-347 (in Russian).
2. V. Bazant, V. Chvaovsky, J. Rathousky, Organosilicon Compounds.
3. L.M. Khananashvili, Ts.N. Vardosanidze, E.G. Markarashvili, N.O. Kupatadze, Synthesis and Properties of Thienylorganosilanes and Siloxanes. J. Polym. Mater, 1992, V. 17, pp. 195-203.

Synthesis and Properties of N-Organosililmethyllactams

Lili K. Janiashvili, Lotari M. Khananashvili, Dali Sh. Akhobadze, Nana G. Manjgaladze, and Antonina A. Surmava

Tbilisi I. Javakhishvili State University

Tbilisi, I. Chavchavadze av. 3, Republic Georgia

Abstract—The reaction of heterofunctional condensation of lactams with chloromethyl derivatives of organosilanes in the presence of catalyst - acceptor of anhydrous K_2CO_3 or Na_2CO_3 has been investigated. The hydrolytic stability of N-dialkyl (aryl) sililmethyllactams in neutral, acid and alkaline media has been studied. The biological activity of synthesized silicon-containing lactams has been investigated. The reaction of hydrolytic polycondensation of N- (methyl dimethoxy-silil methyl) lactams in the presence of catalytic quantity of concentrated hydrochloric acid at room temperature has been conducted.

Introduction

The sililized lactams containing the bonds ≡Si-N= [1-12] are described in literature.

These lactams are biologically active, but hydrolythically unstable.

In [13] it is reported about the synthesis of N- trialkoxysilil methyl - 2- pyrrolidones, obtained by the interaction of N - trimethylsilil - 2- pyrrolidine with chloromethyl - trialkoxysilanes by the following scheme:

However, this method is rather laborius.

EXPERIMENTAL PART

Lactams and triorganochloromethylsilanes were used as the original substances for the synthesis of silicon-containing lactams.

The purity of the original substances was controlled on the gas-liquid chromatograph "LKXM-8MD", phase SKTF-100 (10%) on chromosorb NAW, gas-carrier helium, on the column 2M. PMR spectra were obtained in the spectrometer "Bs-497" of Tesla company (Brno, Chekho-Slovakia) with operating frequency 90MG in solution $CDCl_3$ with interna standard TMC.

RESULTS AND DISCUSSION

Interaction reaction of - 2 - pyrrolidone and - caprolactam with chloromethyltriorganosilanes was conducted at the temperature of 10°C in the presence of anhydrous K_2CO_3 or Na_2CO_3 with molar ratio of reacting substances 1:1 during 4-6 hours in dried toluol medium. The reaction proceeded in one stage by the following scheme:

$$CO-(CH_2)_n-NH+ClCH_2SiR_2R' \xrightarrow[-HCl]{K_2CO_3}$$
$$\rightarrow CO-(CH_2)_n-N-CH_2SiR_2R'$$

where at n=3: $R=CH_3$, $R'=C_2H_5$ (I); C_4H_9 (II) i-C_4H_9(III); C_6H_5 (IV); $R=R'=OCH_3$ (V), OC_2H_5 (VI), OC_4H_9 (VII); $R'=CH_3$, $R=OCH_3$ (VIII) at n=5 $R=CH_3$, $R'=C_2H_5$ (IX); C_3H_7 (X); C_4H_9 (XI); C_6H_5 (XII). $R=C_6H_5$, $R'=CH_3$ (XIII); $R=R'OCH_3$ (XIV); OC_2H_5 (XV); OC_4H_9 (XVI); $R'=CH_3$, $R=OCH_3$ (VXVII).

In the abovementioned conditions the reaction of ω-dodecalactam with chloromethyl triorganosilanes doesn't take place. However, by changing the reaction conditions, that is by conducting the reaction between - dodecalactam and chloromethyl triorganosilanes at the increased temprature (180-200°C) and in the absence of the solvent we succeeded in isolating N - triorganosililmethyl - ω - dodecalactams with satisfactory outputs (34-45%) by the following scheme:

$$CO-(CH_2)_{11}-NH+ClCH_2SiR_2R' \xrightarrow[-HCl]{K_2CO_3}$$
$$\rightarrow CO-(CH_2)_{11}-N-CH_2SiR_2R'+polymer$$

where $R=CH_3$, $R'=C_2H_5$ (XVIII); C_4H_9 (XIX); C_6H_5 (XX); $R=R'=OCH_3$ (XXI); $R'=OCH_3$; $R=OCH_3$ (XXII).

In this reaction, along with N - triorganosililmethyl) - ω - dodecalactams, in all cases rubber-type polymer - poly - ω- dodecane amide is formed, not containing silicon atom in the chain.

Synthesized silicon-containing lactams are colorless, odorless viscous liquids, except N- (trimethoxysililmethyl)- ω - dodecalactam, that is very viscous substance and is crystallized over a few days. All synthesized products are well soluble in various organic solvents.

In table 1 some physico-chemical characteristics, elemental analysis and output of synthesized N-triorganosililmethyllactams are given.

A structure of synthesized N-triorganosililmethyl derivative lactams was established by the help of infrared and PRM spectra.

In infrared spectra of all synthesized compounds the absorption bands of valence vibrations of the bond C-Cl in the region of 780 sm^{-1} are absent. At the same time, in all spectra distinctive absorption bands of deformation vibrations of groups $Si-CH_3$ (1260-1250 sm^{-1}), antisymmetrical and symmetrical valence vibrations of the bonds $Si-CH_2$ in the region of 820-760 and 700-650 sm^{-1}, respectively, are developed. For CO groups, being coupled with nitrogen atom, distinctive absorption bands are found in the region of 1720 sm^{-1} (for N-triorganosilimethyl-2-pyrrolidones), 1680 sm^{-1} (for N-triorganosillilmethyl-ε-caprolactams) and 1650 sm^{-1} (for N-triorganosililmethyl-ω-dodecalactams). Besides, in the spectra of I, IX and XVIII compounds the absorption bands are observed in the region of 1235-1220 sm^{-1} ($Si-C_2H_5$), in II, XI and XIX compounds - 1190sm^{-1} ($Si-C_4H_9$), in IV, XII, XIII and XX compounds - 1429 sm^{-1} ($Si-C_6H_5$), in X compound - 1230 sm^{-1} ($Si-C_3H_7$). Moreover, in VIII, XVII and XXII compounds the absorption bands are observed in the region of 1280 sm^{-1} ($Si-CH_3$) and 1163-1149 sm^{-1} (Si-O-C).

Data of PMR spectra of synthesized N-triorganosilillactams are given in Table 2.

To determine hydrolythic stability of N-dialkyl (aril)-sililmethyllactams, namely, N-dimethyl (phenyl) sililmethyl - 2 - pyrrolidine (IV), N-dimethyl (phenyl) sililmethyl - ε -caprolactam (XII) and N-dimethyl (phenyl) sililmethyl - ω - dodecalactam (XX) they are held at 100°C during 5 hours in neutral; acid and alkaline media. The data on hydroythic stability are given in table 3. These data indicate that sililized lactams with ≡Si-C-N=bond are hydrolythically stable compounds.

Then the reactions of hydrolythic polycondensation of N-alkyl-dialkoxysililmethyllactams were studied.

The investigation of hydrolythic polycondensation of N-(methyldimethoxysililmethyl) lactams showed that depending on reaction's condition this process proceeds ambiguously. So, when this reaction is conducted in methyl alcohol medium in the presence of catalytic quantity of concentrated hydrochloric acid and at room temperature the mixture of oligosiloxanes o the line (73-84%) or cyclic (11-21%) structure is formed. In this case the process proceeds by the scheme:

where at n=3; m=9 (XXIII), p=3 (XXVI) n=5: m=11 (XXIV), p=4 (XXVII)
n=11: m=15 (XXV), p=4 (XXVIII)

The decrease of reaction temperature down to -3÷-5°C leads to the increase of line oligosiloxane output up to 90%.

When hydrolythic polycondensation of N-(methyl-dimethoxysililmethyl) lactams is conducted in ethyl alcohol medium in the presence of catalytic quantity of concentrated hydrochloric acid and at room temperature only six- and eight-term cyclosiloxanes are formed.

The course of the reaction was controlled by the increase of specific viscosity with time (figure 1). As it is seen from figure, hydrolythic polycondensation of N-(methyldimethoxysililmethyl) - ω - dodecalactam proceeds more rapidly that of N-(methyldimethoxysililmethyl) - ε - caprolactam and N-(methyldimethoxysililmethyl) - 2 pyrrolidone and in 90 minutes oligo-dimensional product with maximum specific viscosity (η_{sp}=0.06) is formed, as specific viscosities of the products of hydrolythic polycondensation of N-(methyldimethoxysililmethyl) - ε - caprolactam and N-(methyldimethoxysililmethyl) - 2 pyrrolidine are not higher than 0.035 and 0.028, respectively.

Methyl-lactyl - containing organocyclosiloxanes (XXVI-XXVIII) are colorless transparent liquids, completely soluble in water and chloroform, but oligosiloxanes of line structure are resin type transparent products, well soluble in standard organic solvents.

Synthesis products are identified on the data basis of elemental analysis, molecular mass and infrared spectra.

In infrared spectra of synthesized compounds (XXIII-XXVIII) the intense absorption bands of Si-O-Si bonds are observed in the region of 1030-1010 sm^{-1} for noncyclic organosiloxanes and in the region of 1020-1015 sm^{-1} and 1085-1080 sm^{-1}, being typical of organocyclotri- and organocyclotetra-siloxanes. Characteristic absorption bands of lactam ring are completely retained. In so doing, as differentiated from the spectra of the initial substances (VIII, XVII and XXII) one can observe a significant decrease of the intensity (for XXIII-XXV) and a complete disappearance of the band (for XXVI-XXVIII) in the region of 960 sm^{-1} being related to the asymmetrical valence vibrations of the bond Si-O-C, as evidenced by a complete consumption of methoxy groups.

We've also studied the hydrolythic polycondensation of N-(methyl diethoxysililethylene)- and N-(methyl dibutoxysililethyl)-2-pyrrolidones having been obtained by hydrisililizing of N-vinylpyrrolidone by methyldialkoxysilanes. This process was conducted at room temperature in ethyl alcohol in the presence of catalysts quantities of concentrated hydrochloric acid. The reaction proceeded easily with spontaneous temperature increase up to 60÷70°C. In both cases 8-term organocyclosiloxan is formed by the scheme:

where R=C_2H_5, C_4H_9 (XXIX).

After distilling off the solution and reprecipitating XXIX product from toluol solution by hexane the resin-type product was obtained, that was well soluble in water and chloroform.

In infrared spectra of synthesized compound one can observe a complete disappearance of the absorption band within the region of 960 sm^{-1}, being related to the asymmetrical valence vibrations of the bond Si-O-C. Besides, there is no characteristic of OH group absorption band within the region of 3600-2400 sm^{-1} and characteristic absorption band of the bonds of pyrrolidone ring is completely retained (1720 sm^{-1} for CO).

Physico-chemical data of synthesized compounds are given in Table 4.

The biological activity of synthesized silicon-containing lactams - N-dimethyl-butyl-silil-methyl and N-dimethyl-phenyl-silil-methyl-lactams was investigated.

As test objects the following strains of microorganisms were used: xanthomonas campestris (responsible for the white cabbage bacteriosis), Bacterium tumefaciens (responsible for the vine cancer disease) and Pectobacterium aroideae (responsible for the white cabbage mucous bacteriosis). The results of the investigations are given in table 5.

The analysis of obtained information makes possible to conclude that the substances being tested to a variable degree inhibit the growth of microorganisms and have inhibitory properties.

REFERENCES

1. Pat. 2876209 USA, class 260-46.5. Hurwitz, M.S., Benneville, P.L.
2. Rothe, M., Toth T. Chem. Ber. 99, 3820 (1966).
3. Pat. 2876234 USA, class 260-326,5. Hurwitz, M.S. Benneville P.L.
4. Rühlmann, H.H. Rupprich. B., S. Lieb. Ann. Chem. 686. 226 (1965).
5. Andrianov, K.A., Nogaideli, A.I., Khananashvili L.M., Akhobadze D.Sh., Tkeshelashvili R.Sh. "Izv. AN SSSR", Chem. 5, 1157 (1970).
6. Andrianov K.A., Nogaideli A.I., Akhobadze D.Sh., Khananashvili L.M., Tkeshelashvili R.Sh. "Izv. AN SSSR", Chem. 5, 1150 (1972).
7. Andrianov K.A., Nogaideli A.I., Khananashvili L.M., Akhobadze D.Sh., Vardosanidze Tz.N. "Soobshch. AN GSSR', 63, 3, 605 (1971).
8. Nogaideli A.I., Akhobadze D.Sh., Khanashvili L.M., Jibgashvili G.G. "Soobshch. AN GSSR", 67, 2, 345 (1972).
9. Bergman H.S., Mayhoffer R., Hartwig H. Arch. Pharm. 319, 3, 203 (1986).
10. Kramarova E.P., Anisimova N.A., Baukov Yu.I. "Jurn. Obshch. chem." 46, 7, 1658 (1976).
11. Woodbury R.P., Rathke M.W.J. Org. Cem. 43,5, 881 (1978).
12. Kricheldorf H.R. Leppert E. Makromol. Chem. 158, 223 (1972).
13. Kramaroa E.P. Author's abstract of Cand. Chem. Sc. 1968, p. 5-6.

On Mechanism of Liquid Water Isothermic Segregation in a Medium of Highly Elastic Hydrophobic Polymers (Using the Example of Polymethyl Methacrylate)

Yu.A. Mikheev, L.S. Pustoshnaya, and G.E. Zaikov
Institute of Biochemical Physics, Russian Academy of Sciences
4, Kosygin st., 117334, Moscow

Deglassifying the films of plasticized polymethyl methacrylate (PMMA) in heated water causes the spontaneous segregation of micrometer water drops and over in radius and is accompanied by a change in the film sizes. This process is impossible in structureless hydrophobic systems and points to the structural organization of the polymer chains, which are responsible for adsorption of a certain amount of molecular water by hydrophobic PMMA. From a thermodynamic standpoint, "structural hydrophily" is of entropic nature and is provided by the micropore generation due to the fluctuation. The micropores are stabilized owing to the capture of water molecules. When the polymer transforms to the highly elastic state, zones of a dynamically active microporous sponge are formed in it with the moisture participation. In these zones, the fluctuation results in the formation and destruction of hydrophobic slits, which ensures the appearance of water interlayers and drops of critical size that merge into the stable water phase. A thermodynamic model of the process is considered.

Water diffusion in highly elastic hydrophoblic polymers is complicated by the segregation of water drops [1-3]. This phenomenon is of practical interest because of its influence on the exploitation properties of polymeric materials. However, a theory of the water phase formation under such conditions has not been completed. All the investigations performed in this line are based on the concept of interaction between the diffusing water molecules and water-soluble impurities, which are present in the materials.

At the same time, the above phenomenon is typical for many polymers that differ in technology of their production and processing (natural and synthetic rubbers, polyolefins, polyvinyl chloride, polycarbonate, polysulfone, polyvinyl acetate, and so on [1-3]). This fact is in rather poor agreement with the presence of water-soluble impurities in all the polymers and suggests that the more common mechanism of the phenomenon exists.

In this work, the formation of the drop water phase is studied by the example of plasticized hydrophobic PMMA. The water-soluble impurities are not necessary for the realization of this process, but the high molecular-segmental mobility is required.

EXPERIMENTAL

Commercial PMMA (M_w=135 000, M_w/M_n=2.0, T_g=380 K), which was produced by emulsion polymerization of MMA in the presence of lauryl peroxide and contained about 1% of residual MMA, was used both in unpurified and purified forms. The purification was conducted through the polymer precipitation from a methylene chloride solution by methanol followed by additional extraction by methanol in a Soxhlet apparatus for 6 hours. The 5-35 μm thick films were prepared from a PMMA solution in distilled methylene chloride (MC) at the polish glass surface by their exfoliation with water upon MC evaporation. Then, the films were kept in a vacuum for 24 hours. In order to accelerate the structural relaxation, a portion of the films was annealed in a vacuum at 343-353 K for eight hours.

For PMMA plasticization, spectroscopically pure dibutyl phtalate (DBP) was used, its concentration in MC solutions being 2,5, 6.0, 10.0, 20.0, and 30.0%. Dibutyl phtalate and PMMA have virtually the same solubility parameters equal to 19.3 and 19.0 $(J/cm^3)0.5$, respectively, whereas that for water equals to 47.0 $(J/cm^3)0.5$ [4]. The above fact points to identical hydrophoby of both components; therefore, DBP cannot be considered a water-soluble impurity.

To determine the amount of liquid water segregated, the sizes and number of water drops, the method of the turbidity spectra was used, which is generally applied for such purposes [1,3,5,6]. The method is based on the difference in the indexes of refraction for water and polymer. These indexes are equal in magnitude for PMMA and DBP (1.49).

The turbidity spectra was recorded using the UV Vis Specord instrument. The film strips of size 10×30 mm were placed in spectroscopic cells, which were filled with distilled water and were heated in a thermostat. The volume content of water drop phase Φ% and the average radius of the drops R were calculated by formulas $\Phi=\tau/g$ and $R=\alpha\lambda/2\pi\mu 0$, where $\tau=2.3D/1$ is the turbidity of the 1 cm thick film, D is the optical density at λ=546 nm, μ_0=1.33 is the index of refraction for water, α and g are the parameters determined from the calibration graphs given in [5,6]. The n index, which is required for the use of these graphs, was found from logarithmic anamorphoses of the D=const λ^n relationships in the range 400<λ>700. The relative index of refraction $m=\mu/\mu_0$ and the index of refraction μ for optically transparent PMMA saturated by water (at the water volume fraction of 0.025) are taken equal to 1.115 and 1.486, respectively [2,7]. The radius of voids, which are left upon drying out the turbid samples, was calculated by formula $R=\rho\lambda/2\pi\mu$ (λ=500 nm, μ=1.5, ρ was determined from the calibration plot of n versus ρ [8,9].

RESULTS

The turbidity spectra for the films of unpurified PMMA are observed at a temperature of water above 343 K. Figure 1 shows these spectra, and Figure 2 shows the corresponding Φ(t) curves. The turbidity increases as water temperature rises. This process is sensitive to the structural relaxation proceeding in the polymer. Thus, the freshly prepared films reach fast the limiting turbidity (Figure 2, curves 2a, 5-7). However, time of the attainment of the limiting value increases on prolonged keeping the films. If such samples were preliminary annealed at 343 K, these virtually do not become turbid in water at the same temperature (Figures 1 and 2, curves 1a and 1b), but lose their initial activity (Figure 1, curves 5 and Figure 2, curves 2a and 5). At elevated temperatures this process proceeds more slowly in the annealed films (Figure 2, curves 3 and 4) compared to the unannealed ones (Figure 2, curves 6 and 7).

The decrease in the thickness of films prepared from unpurified PMMA causes a disproportionately sharp reduction of the limiting turbidity. Thus, Figure 1 (curve 2) shows the insignificant turbidity for the 10 μm thick film at 368 K. This value is rather far from 0.28 the value that to e expected from the 4,5 and 6 curves observed at 363, 353, and 368 K, respectively, for the 35 μm thick films. The above peculiarity is connected with fast removing the plasticizer remains (MMA and MC) from the thin films. With the dry anneal of samples, as distinct from the water one, the MMA and MC remains are poor removed from the films. Thus, the IR absorption spectra show that the bands at 700 and 735 cm^{-1}, which are characteristic of MC, virtually do not change even after the additional film treatment in a vacuum for 10 hours. The above bands hardly decreases after the anneal, but disappear completely when the films are heated in water for a short time.

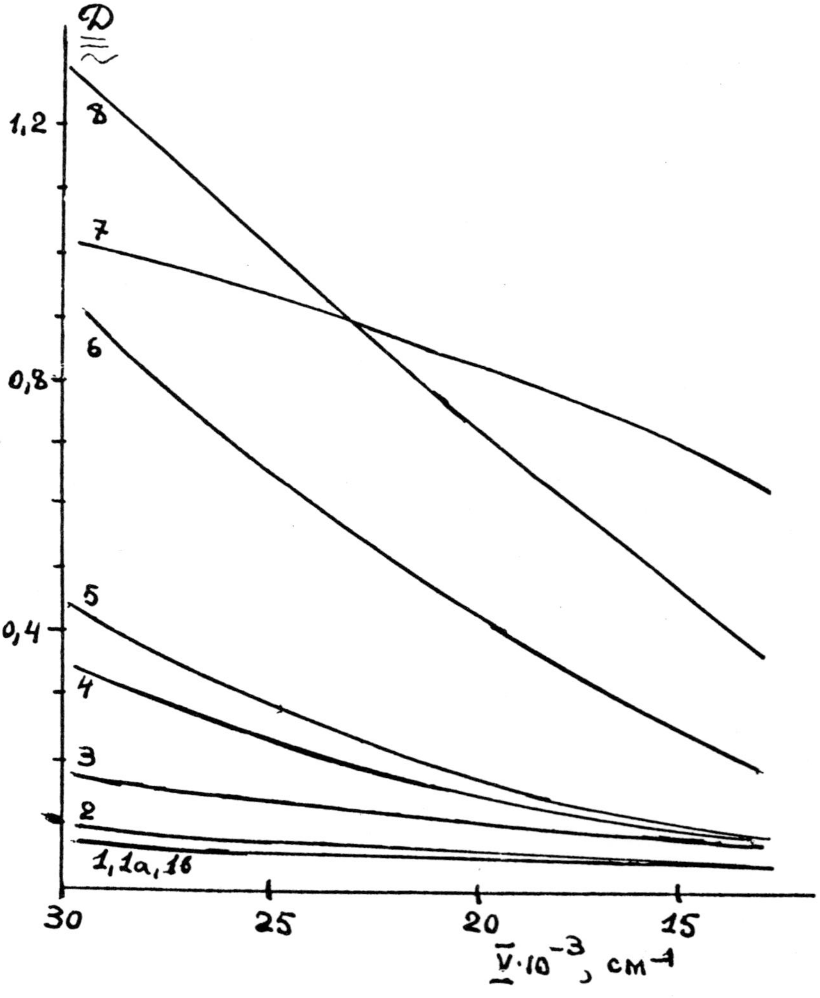

Figure 1. The turbidity spectra of the PMMA films: arbitrary thick (1), 10 μm (2) and 35 μm (1a, 1b, 3-8) thick after their heating at 343 (1a, 7), 353 (1b, 5), 363 (4), 368 (2,6) and 372 K (3) for 200 (1a,1b), 20 and 100 (2), 5 and 50 (3), 10 (4), 20 and 200 (5), 10 (6), and 110 min (7) prepared from the unpurified (1a, 1b, 2, 4-8) and reprecipitated (3) polymers and from that containing 27% of DBP (7) with the preliminary anneal of the films (1a, 2b, 3) and without the anneal (2, 4-8). The spectra are obtained in an optical cell filled with water (1-7) and in air (8), the latter spectrum is recorded after drying the film exhibiting spectrum (4).

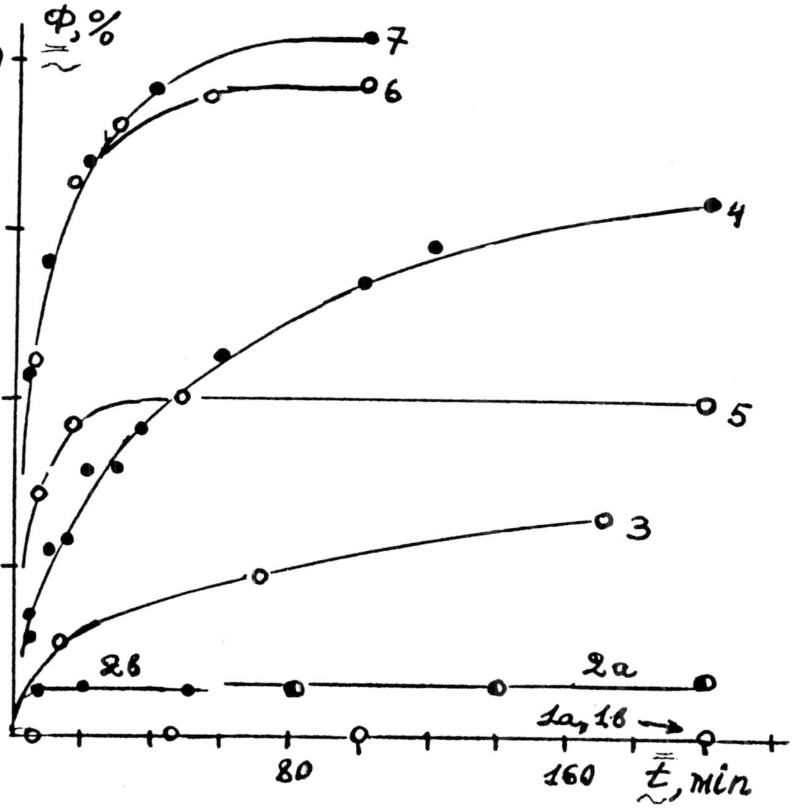

Figure 2. The dependence of the volume content of drop water in the polymer on heating time for the 35 μm thick films prepared: from unpurified PMMA with the dry anneal of the films (1a, 1b, 3,4) and without the anneal (2a, 5-7) and from reprecipitated PMMA with the dry film anneal (2b) at temperatures of 343 (1a, 2a), 353 (1b, 5), 363 (3,6), 368 (4,7), and 372 K (2b).

The role of residual placticization is minimum for monomer-free samples. The turbidity of such samples is insignificant, and the process stops fast (Figure 1, curve 3 and Figure 2, curve 2b) owing to the rapid removal of the MC remains.

As distinct from the above situation, the stimulating effect of the greater sized DBP molecules on the process is retained for any thick films for a long time over a wide range of temperatures. In particular, curve 7 in Figure 1 was obtained at the DBP content of 27%, at 343 K in a time of 100 min, whereas the samples did not become turbid in the absence of DBP (Figure 1, curve 1a).

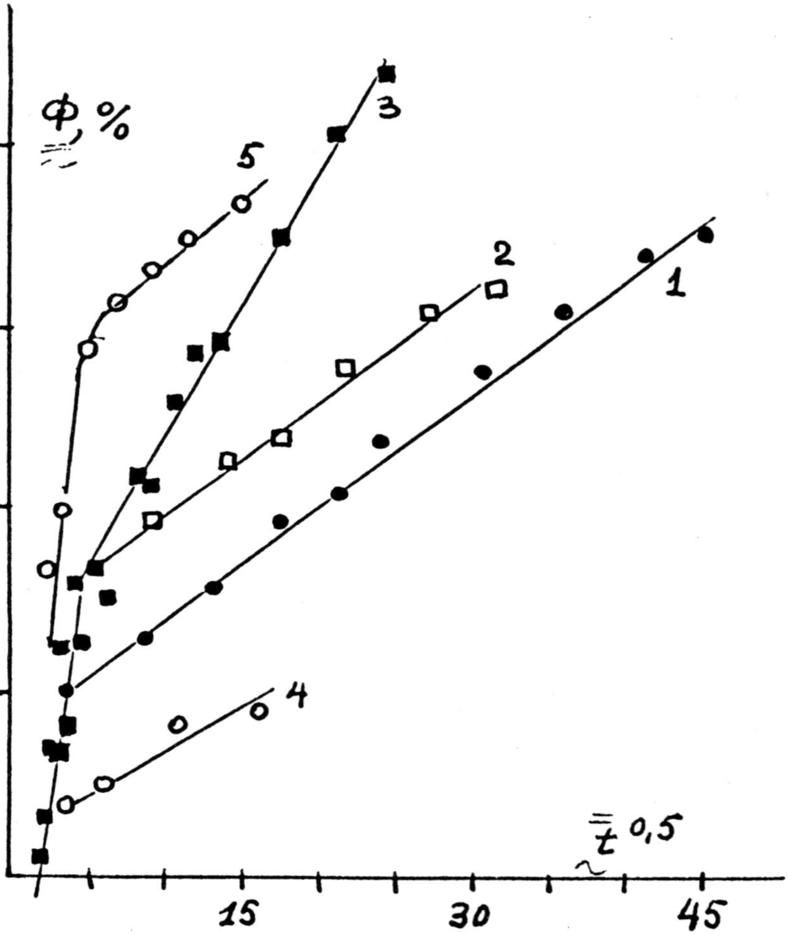

Figure 3. The curves of the accumulation of drop-liquid water in plasticized PMMA at the DBP content of 6(1), 10 (2,3), and 30% (4,5) for the films of 18 (1), 16 (2), 25 (3), 5(4) and 13 μm (5) in thickness.

Figure 3 shows in coordinates Φ, $t^{0.5}$ the curves of the drop water accumulation at 363 K in the samples, containing 6-30% of DBP. The process is accompanied by a small loss of DBP only (no more than 10% for 1.5 hour at the DBP content of 30%). The shape of the curves in Figure 3 points to the presence of two kinetic stages in the process - the fast and the slow ones. These stages cannot be given a common quantitative description, because the different Φ values correspond to the same instants of time at the same DBP concentrations.

The process under study is accompanied by a characteristic change in the sample sizes: their thickness increases in parallel with a decrease in the length and width. Thickening is especially great (up to 50%) when the film edges is not rigidly fixed. The rigid fixation decreases the turbidity rate. However, the internal stress arised causes often an abrupt increase of the turbidity in the course of the experiment owing to drawing the film edges from the fixing frame. Thus, the data in Figure 3 reflect not only the internal properties of the samples, but the action of the outside mechanical stops of the deformation.

Figure 4 shows the temperature dependence of Φ for liquid water, which is accumulated for 15 min at the various DBP contents. The 1-5 points refer to the DBP-free films, of which the 1-3 ones are obtained at 368 and 372 K and correspond to the limiting Φ value, because these characterize reprecipitated PMMA, losing fast the MC remains. The 4 and 5 points characterize annealed and unanealled films, respectively, which were tested at various intervals during their keeping at room temperature. As is seen, PMMA, whose internal structure is not completely relaxed, displays a higher activity. Thus, about 2 and 0.2-0.3% of water is segregated for 15 min in the freshly prepared films and in the annealed ones, respectively..

The relaxation factor is in evidence at the low DBP content as well. As an example, we refer to the 6,7 and 9,10 curves that were obtained at the DBP content of 2.5 and 10.0%, respectively. Note that every point in the 6 and 9 curves was obtained using an individual sample of the freshly prepared film, and those in the 7 and 10 curves were obtained without the sample exchange, when the heat action was accumulated. The curve 8 corresponds to the film aged enough, containing 6% of DBP. The curves 11 and 12, which were obtained for the samples with 20 and 30% of DBP, respectively, are identical for the freshly prepared and aged films.

The date presented in Figure 4 show that drop water appears in plasticized samples at glass transition temperatures, T_g, which are determined for analogous samples by the DSC method [4] and equal to 355-360, 345-350, 333-338, 315, and 300 K at the DBP content of 2.5, 6.0, 10.0, 20.0, and 30.0%, respectively. Thus, DBP displaying the hydrophoby identical with that of PMMA, causes the segregation of liquid water drops owing to deglassifying the polymer. It is notable that as the DBP content increases, and the T_g value is correspondingly reduced, the slope of the straight-line portions of the 8-12 curves decreases, i.e., the temperature sensitivity of the developed process is diminished. This looks as a somewhat unusual counteraction between the factor of plasticization and that of heating, although the above factors would be expected to act on the activation of the molecular-segmental mobility in the same direction.

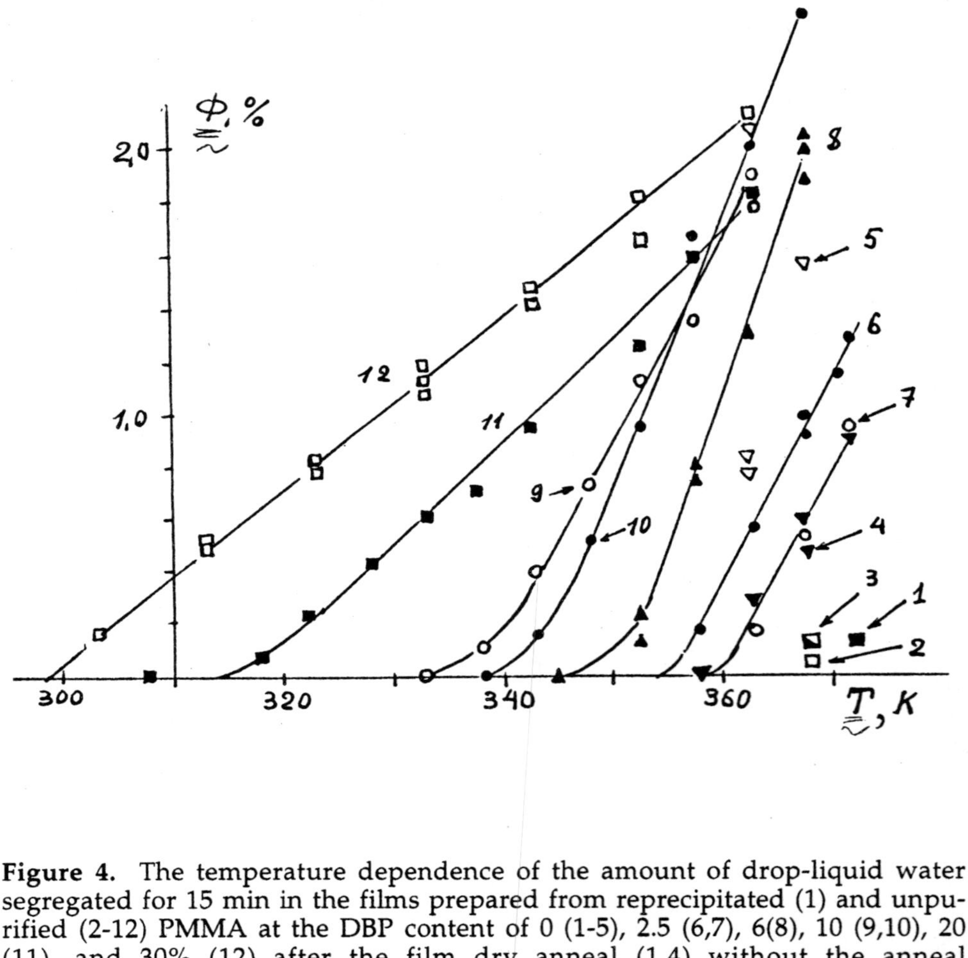

Figure 4. The temperature dependence of the amount of drop-liquid water segregated for 15 min in the films prepared from reprecipitated (1) and unpurified (2-12) PMMA at the DBP content of 0 (1-5), 2.5 (6,7), 6(8), 10 (9,10), 20 (11), and 30% (12) after the film dry anneal (1,4) without the anneal (2,3,5,6,8,9,11,12) and when a temperature of the single sample was stepwise increased (7,10). The thickness of the films is: 10(2), 13 (12), 17(3) and 30 μm (5-11).

The following may be mentioned regarding the average radius of water drops R and their number N in unit volume. In the absence of DBP, the R value is constant from start to finish of the process and is the same for unannealed and annealed films: 210±20 and 310±20 nm, respectively, at 353-363 and 368-372 K. In the initial stage of the process, the N number in the unannealed polymer grows more sharply compared to that in the annealed one. For example, at 368 K, N equals to the following values: !.0 10^{11} and 0.1 10^{11} cm^{-3} (within 5 min), 1.8 10^{11} and 0.9 10^{11} cm^{-3} (within 100 min) for the annealed and unannealed polymers, respectively (Figure 2, curves 7 and 4). At the same instants of time at 363 K, the corresponding N values equal to 2.8 10^{11} and 0.5 10^{11} cm^{-3}, 5.0 10^{11} and 1.4 10^{11} cm^{-3} (Figure 2, curves 6 and 3). Thus, a change in the internal structure, which is caused by the dry anneal, decreases the growth of the drop number.

In the presence of DBP, the average radius of drops in PMMA increases during the process in the ranges 300-450, 400-700 and 300-1000 nm at the DBP content of 2.5, 6.0 (10.0) and 20.0 (30.0)%, respectively That is, the R value exhibit the variability of its magnitude (as the Φ factor does) due to the influence of mechanical stops on the elastic-viscous deformation of the material.

The considerable resemblance of the liquid water segregation in deglassified PMMA to the corresponding process in elastomers [3] should be noted. However, turbid elastomers become transparent reversibly as a result of their drying, whereas this process in PMMA is hindered by glass transition. On the contrary, the turbidity of dry PMMA when increases when the DBP content is low or in the absence of the latter (Figure 1, curves 4 and 8). This phenomenon results from an increase in light scattering due to the replacement of liquid water for air. The calculations by the method described in [8,9] showed that the average radius of voids in dry DBP-free films equals to that of segregating water drops. Repeated wetting the dry turbid films brings about their initial turbidity (Figure 1, curves 4), the drop radius being unchanged. The above result indicates that interaction between the PMMA transparent films and water is not limited to filling the available voids and is connected with the displacement of the polymer matter from the zones of growing drops. This result is also indicative of the absence of dissolved macromolecules in the drops.

The initial transmittance of the dry turbid films is restored by heating only. The films become transparent at the marked rate at temperatures close to the T_g value (380K), but the process slows down rapidly, because an additional rise of a temperature is required. At 393 K, the process has completed in about 0.5 hour. Processing the turbidity spectra by the method [8,9] showed that clarifying the films proceeded through reducing the void number, the R value being decreases insignificantly. That is, this process occurs not at the expense of viscous flowing the material into the voids, but owing to defrosting the stored highly elastic deformation and to fast collapse of individual caverns. Indeed, a prolonged decrease in the R value for many tens of hours would correspond to the mechanism of the viscous flow because of very high viscosity of the polymer [10].

DISCUSSION

According to the above considerations the high molecular-segmental mobility in PMMA ensures isothermic interaction between water and polymer in terms of the following scheme:

water of an outer bulk $\xrightarrow{I}$ *molecular surface water* $\xrightarrow{II}$ *molecular dissolı water* $\xrightarrow{III}$ *drop liquid water.*

An important point is that the phase transition of molecular water diffusing in the polymer is not connected with water-soluble impurities or with filling the initially existed voids and occurs at the expense of a mechanical work.

The diffusion coefficient for the water molecules in PMMA at T_g is known to be equal to 3.3 10^{-8} cm^2/s and to increase as a temperature rises [11]. Time, $t=x^2/6\ \alpha$ [12], of saturation of the films 40 μm (l) in thickness (x=0.5l) by molecular water should be less than 20 s. Therefore, the segregation of drop water (stage III), which is continued for many hours, is not limited by diffusion stage (II) and is determined by the specific structural properties of the polymer. Note that in a homogeneous structureless medium, such a process is impossible due to the absence of medium supersaturation by dissolved water under isothermic conditions and to the contradiction with the second thermodynamic low [13].

As was mentioned above, this problem is considered in various publications in terms of the concept of the low degree of structure of a medium. Therefore, it is always suggested the formation of osmotic cells involving water-soluble impurities which are heterogeneously distributed in the structureless medium [1-3, 13]. This assumption is unsuitable for our case, sending us in search of the reason for the phenomenon observed in the peculiarities of the supermolecular structure that ensures the occurrence of stage III via a certain operating element.

Obviously, the structural element - generator of the initial water drops - should exhibit the opposite properties and operate in the following way. First, it will accumulate a needed amount of water molecules (in spite of hydrophoby inherent in macromolecules) and then will segregate accumulated water as drops, realizing the hydrophoby of the macromolecules. To the above scheme there corresponds a microporous sponge formed from the dynamically mobile polymer chains. The affinity of the sponge to water depends on the appearance of fluctuation micropores. It is precisely these micropores that sorb water in hydrophobic highly elastic polymers [14].

The work of the generator of water drops can be described in terms of the concept, according to which the main structural element in noncrystalline polymers is a "superglobule" which is formed from a great number of macromolecules [15]. The PMMA globules observed in the electron-microscopic photographs are 10-15 nm in diameter [16, 17]. According to [18,19] every globule is insulated by a shell of the tightly packed segments of polymer chains. Tending to have the conformation of statistical tangles in the globule bulk, the chains undergo more or less strong stretching in the radial direction of the globule, thus redistributing load on the shell.

For glassy PMMA and PS, the stretch of macromolecular tangles is proved by the methods of small-angle X-ray scattering and neutron scattering [20-22] (their radii are more than those of tangles in a Θ-solvent by 10-20%). The above stretch is also demonstrated by a jump of an internal pressure at a temperature close to T_g, which is connected with the storage of the energy of the chain elastic tension as much as 15-30 cal/cm3 in PMMA and PS glasses [23].

Glassy PMMA falls in a category of nonporous sorbents [24]. However, mechanical processing the polymer in a ball mill at 77 K imparts it the sorption capacity with respect to nitrogen that is equal to that of activated carbon [25], suggested that the expended work goes into loosening the globule filling to the state of the rather developed sponge.

The typical for usual PMMA fraction of the free volume, providing the density fluctuation at T_g, equals to 0.019 (the method of small-angle X-ray scattering [26]) or to 0.021 (the procedure of hydrostatic compression [27]). At the same time, the relationships of water vapor sorption indicate that the structural voids stored in PMMA are not primarily filled. Thus, the bulks of polymer and water are approximately additive until the water content attains 2 wt.% ($P/Ps \approx 0.95$) [7], and the heat of sorption equals to zero. At the relative moisture $P/Ps > 0.95$, a deviation from the Henry' low is observed owing to the appearance of internal structure surfaces in PMMA, which separate water clasters consisting of some molecules and molecular chains with the hydrogen bonds [2]. It appears to be advantageous for the chain sponge to physically bind a certain amount of water molecules and associates. This process results in a change in the chain conformation and in a decrease of the elastic load in the sponge. The entire energy of such interaction goes into the work directed against the structural forces. This phenomenon makes itself evident in the heat effect equal to zero and in the formation of a colloidal solution of osmosis-bonded water [28].

Let us consider the system, containing a constant number of particles, using the method of thermodynamic potentials. From the basic thermodynamic relationship $TdS=dU+PdV+\delta W$, where δW is the elementary work against structural forces, the following expressions for the Gibbs' energy and the enthalpy, respectively, can be written: $dG=-SdT+VdP-\delta W$ and $dH=d(U+PV)=TdS+VdP-\delta W$. Hence we obtain the isobaric-isothermic process $-dG_{P,T}=\delta W$. Because in the case considered $dH=0$ and $dU=0$ ($PdV=0$ due to additivity of the bulks of the components), we have $-dG_{P,T}=TdS=\delta W$ or $-\Delta G_{P,T}=T\Delta S=W$. Thus, the driving force of the consumption of molecular and claster water by a hydrophobic polymer when the sorption heat equals to zero is a decrease in the Gibbs' energy caused by an increase in the entropy of the system due to the growth of the entropy of the polymer-chain network and to mixing the components.

Based on the thermodynamic theory of vacancy solutions universal for nonporous, wide-porous, and microporous sorbents [30,31], one can consider that moistened intraglobular filling is turned to a sponge with an ideal vacancy solution over the interval, wherein the Henry's low is valid ($P/P_S \leq 0.95$). As clasters and chains of associated water molecules appear, the vacancy solution ceases to be ideal. In accordance with the concept developed in [30, 31], a binary solution, comprising an adsorbed substance and adsorption vacancies, is always at osmotic pressure in osmotice equilibrium with the adsorption vacancies.

In other words, the globule filling of a hydrophobic polymer carries a structural (entropic) "water-soluble" component produced by the elastic tension and by the flexibility of the polymer chains. Adsorption vacancies - fluctuation micropores - act as such a component. Due to the capture of water molecules and clasters, the micropores are stabilized as elementary osmotic cells and create the osmotice pressure in the moistened sponge. However, the factor of vacancy-entropic hydrophily by itself cannot ensure the integration of water, which is colloidally dispersed in the body of the sponge, into the drops of the stabile liquid phase. The realization of this phase transition is impossible without supersaturation of the sponge medium by water.

The mechanism of the supersaturation comes into effect with the availability of a high molecular-segmental mobility of the polymer ($T \geq T_g$) if the sponge structure enables the cooperation of the chains segment movements under the water action. In circumstances where fluctuation cooperative oscillations of hydrophobic segments occur, slits broad enough can open and collapse in the sponge matrix. These slits accumulate water in the amount required for the appearance of the water interlayers and drops due to the fluctuation. As this take place, the factor of supersaturation appears as a result of the manifestation of hydrophobic forces, realizing via a disjoining pressure π [32].

The thermodynamic parameter π connected with the colloidal state of a substance constitutes a variety of components [32]. For this consideration, the negative component resulted from the forces of the molecular attraction of the slit walls and the structural one connected with the structure of the water interlayer are of importance. The latter component is different in its structure for hydrophobic and hydrophilic slits, the wall attraction to one another and the wall repalsion, respectively, being inherent in the first and second cases. The point is that the water density is reduced near the hydrophobic walls, and that is enhanced in the bulk between the walls. In this case, the attraction force of the walls exceeds considerably that realizing through the homogeneous water interlayer. On the contrary, in hydrophilic slits, water is strongly linked to the walls, forming thick adsorptive layers, mutual overlapping of which ensures a total positive disjoining pressure [32].

According to [32], in the general case of thermodynamic equilibrium, expression (1) takes place

$$\pi = P - P_V \quad (1)$$

where P and P_V, respectively, are pressures at the interlayer surface and in the bulk, which serves as the interlayer extension. However, with overlapping the ordered adsorptive layers, the possibility of the drop formation from unbonded water is excluded in hydrophilic slits (in particular, the drops do not form in a wood with the slits attaining a width of 10 nm [28]). For such layers, which are in equilibrium with vapor, the π value is determined by the expression identical to that for osmotic pressure P_{os} [32, 33]:

$$\pi=P_{os}=RT\ln(P_s/P)/V_m\geq 0,$$

where V_m is the molar volume of film water. The relative vapor pressure $(P/P_s)\leq 1$ points to the absence of slit supersaturation by unbonded water. If a portion of water molecules unites accidentally into a germ drop, then the latter, exhibiting an elevated vapor pressure $P/P_s>1$ [33], will immediately evaporate into the layer of bonded water.

In the case mentioned above, the subtrahend in equation (1) disappears ($P_v=0$), and the thermodynamic equilibrium is realized under conditions of balance between the external and internal pressures, which are identical to the swelling pressure: $\pi=P_{os}=P$.

In accordance with these considerations, the polymer-chain sponge formed from hyrophilic slits enter in quasi-equilibrium interaction with water and should swell uniformly in any arbitrarily taken region.

As distinct from the above situation, entropic hydrophily of a hydrophobic sponge is realized in the form of the identity of π with P_{os} only on a limited scale of swelling, which is nonuniform in various zones of the matrix. The swelling is limited due to the fact that opening the hydrophobic slits results in negative disjoining. The latter causes the reverse collapse of slits under conditions of the high structural dynamics and in doing so prepares the new alteration of the sign of the disjoining pressure. The quasi-equilibrium state of the hydrophobic slit can be ensured by a stretching pressure that is equivalent to $-\pi$. As this takes place, the following conditions will be fulfilled for the water interlayer: $\pi=-P_{os}=-RT\ln(P_s/V_m<0$ and $P/P_s>1$ (the latter points to slit supersaturation by unbonded water). Such a situation is unstable, but that is possible, when stretching of a zone is temporarily stabilized owing to compressing a material in the adjacent one.

The above combination of the polymer chain displacements is unlikely in the bulk of the uniformly swelling hydrophilic sponge; however, that is realized in the hydrophobic one owing to the appearance of waves of medium expansion and compression in the neighboring zones under the interchange of the π sign in opposed phases of the oscillation. Accidently produced synchronization of such structural oscillations can result in the fluctuations of capillary water when, in certain sponge regions, drawing off of the water interlayers from the neighboring zones that are in compression occurs in the expansion phase, and then the fluctuation water drops segregate during the next compression phase.

In the instant of the expansion of an individual slit, containing the water interlayer, the cooperative movement of the polymer links is capable to result in complete releasing the pressure on water, so that the slit is converted into a sphere of gathering the water films from a zone being in compression.

In the latter zone, the total pressure exerted on the water interlayer can be doubled in the limiting case owing to the sign interchange of the external (relative to the interlayer) pressure, which counteracted the collapse pressure $-\pi$ prior to the above instant.

A new metastable state of the mechanical equilibrium is characteristic of the situation being considered. In this state, the doubled pressure on the water interlayer, communicating with a fluctuation drop, is balanced by the opposing pressure exerted by a drop (the latter pressure is generated by the force of interfacial tension). This state is destroyed in the course of structural fluctuations and either returns to the molecular-claster dispersion or provides merging the fluctuation drops into the stable liquid phase.

The realization of the fluctuation acts does not change the internal energy of the system (dU=0). Therefore, it follows from the expression for the first thermodynamic low $\delta Q = dU + PdV$ that the fluctuation work should be provided by the isothermic consumption of the heat, when the drops are formed, and by the heat liberation, when the reverse dispersion occurs to form clasters and interlayers.

According to the fluctuation theory of phase transitions [33], the formation of drops of critical size is connected with an increase in the Gibb's energy (ΔG) by one third their surface energy. This excess should be compensated by the isothermic work of a hydrophobic sponge, which is directed to maintaining a critical drop at the metastable equilibrium. Taking into account the above considerations, we can write the following equation:

$$\Delta G_c = (4/3)\pi r_c^2 \sigma_c = P_c V_c = \Delta Q_c,$$

where r_C and V_C are the radius and the volume of the critical drop, respectively, σ_C is interphacial tension at the polymer-critical drop boundary. From this equation, we have

$$P_c = -2\pi = 2P_{os} = \sigma_c / r_c. \qquad (2)$$

The P_{OS} value can be estimated from the isotherm of water sorption by PMMA [7], if in accordance with the concept of vacancy solutions [30], P_{OS} is expressed via Gibbs' integral

$$P_{os} = \int_0^{(P/P_s)=1} C(P/P_s)\, d\mu,$$

where $C(P/P_S)$ is the equilibrium molar concentration of water sorbed by the polymer, μ is the chemical potential of the adsorbate. Hence, we arrive at

$$P_{os} = RT\int_{0}^{(P/P_s)=1} C(P/P_s)d\ln(P/P_s) \tag{3}$$

As was mentioned above, the isotherm of sorption given in [7] does not depend on T and deviates only slightly from the Henry's low in the range of the P/P_s values from 0.95 to 1. Therefore, we can neglect by the fact that a vacancy solution forming in PMMA is not ideal and express the isotherm by the equation of a straight line

$$C = C_1(P/P_s), \tag{4}$$

where C_1=1.5 mol/l (0.023 wt.%) is the value of water concentration obtained by extrapolation to the value $(P/P_s)=1$. Integration of equation (3) in view of (4) gives the formula $P_{os}=C_1RT=1{,}25\times10^4T$, which coincides with the van't Hoff's equation for an ideal liquid solution. According to this formula, the appraised values of the osmotic pressures are equal to 3.78, 4.15 and 4.53 MPa at 303, 333 and 363 K, respectively.

Using the well-known expression for the disjoining pressure, $\pi=A/h^3$ [28] (where $A=5\times10^{-14}$ erg and h is the thickness of the water interlayer), we found the average value of h for fluctuating slits, which lies in the range 1.10-1.03 nm at 303-363 K. This value corresponds approximately to four water molecules.

The radius of a fluctuating drop of critical size can be estimated provided that the interfacial tension σ_c depends on the surface curvature $\sigma_c=\sigma/(1=+21/r_c)$[33], where 1=0.5 nm is the thickness of a surface layer of liquid water, σ is the interfacial tension calculated by the Owence-Vendt's formula [12] based on the surface tension values for PMMA and water [34]. The calculated σ values are equal to 26.3, 22.8, and 19.0 dyne/cm at 303, 333, and 363 K, respectively. According to equation (2), we have $P_c=2P_{os}=\sigma/(r_c+21)$; hence, the r_c values are equal to 2.50, 1.75 and 1.10 nm at 303, 333 and 363 K, respectively. Note that the values found should be related to deglassified PMMA. At the temperatures used this is achieved by applying DBP as plasticizer; DBP should not influence the σ value, because its solubility parameter is identical with that for PMMA [12].

The critical drops formed take up a part of the free volume, the fraction of which in PMMA (0.021 at Tg) ensures the formation of caverns, 2.7 and 6.9 nm in radii, inside the globules of diameters 20 and 50 nm, respectively. These values are large enough for the spatial redistribution of polymer chains without accumulation of additional loads on them.

As a result of integration of the critical drops in the course of the structural fluctuations, the stable water drops arise. These drops continue to grow overcoming the elastic viscous resistance of the polymer matrix, formula (2) being not fulfilled. At this stage, a new quasi-equilibrium state is established, when an excess (Laplace) pressure P_L (generated by the interface forces) on the drop is balanced by the pressure P_c, which is favorable for an increase in the interface:

$$P_c=P_L=2\sigma_e/r, \tag{5}$$

where r is the radius of the growing drop, σ_e is the interfacial tension at the boundary of the elastically compressed polymer matrix. The σ_e value can exceed considerably the σ value that is characteristic of the plane surface [33], which is not deformed. In the case considered, the drop radius determined at the early stage of the water segregation, when elastic-deformation phase of the process prevails, is 300 nm. Using formula (5), we can find the σ_e values, which are equal to 43σ, 54σ, and 70σ at 303, 333, and 363 K, respectively.

The isothermic phase-forming work of the highly elastic polymer-chain sponge reacts to mechanic obstacles with a high sensitivity owing to its fluctuation origin. In particular, this is true regarding the obstacles caused by the force of mechanical fixation of films or by additional ordering the internal structure due to the sample anneal.

Moreover, the growing drops by themselves decrease the intensity of the fluctuations and the rate of the liquid segregation because of elastic loads on the matrix and deformation of the latter. Therefore, there are two stages of the process in time - the phase of essentially elastic deformation changes for the phase of viscous flowing. Figure 3 demonstrates the influence of the above change in the deformation regimes on the rate of the liquid water segregation.

Elastic compression of the polymer containing a great number of water drops differs little from the compression by a hydrostatic pressure, P_h. This fact allows us to estimate the bulk modulus of elasticity (usually determined by formula $K_h=P_h/(\Delta V/V_0)$ where $\Delta V=V_h-V_0$; V_0 and V_h are the bodies of the polymer in the initial and compressed states, respectively) via the volume fraction of liquid water Φ_e, which is segregated at the stage of elastic deformation on the matrix

$$K_h=P_c/\Phi_e. \tag{6}$$

The Φ_e values found from Figure 3 by extrapolation of the straight-line portions, corresponding to the second stage of the process, to the ordinate axis have the scatter from 0.013 to 0.036. This scatter appears to reflect both the difference in the internal structure and in the force of mechanical fixation of the samples. Consequently, the calculation by formula (6) results in the K_h values for plasticized PMMA at 363 K ranging from 700 to 250 MPa (for unplasticized PMMA, $K_h \approx 2000$ MPa [27]).

The appraised value of the Young's modulus obtained by formula $E=3K_h(1-2\nu)$ [12] (where $\nu=0.495$ is the Poisson ratio characterizing the rubber-like state) lies in the range 21 - 7.5 MPa. Another method of the Young's modulus estimation, taking into account the size of water drops by formula $E=6P_c/(5-4/\lambda-1/\lambda^4)$ [3] (where the coefficient $\lambda=100$ corresponds to the degree of increasing the radius of a cavern driven apart by water from 3 to 300 nm), results in the value of $E \approx 11$ MPa. These results obtained on the basis of the water phase characteristics confirm, on the one hand, that plasticized PMMA is among rubber-like materials and that the interpretation of the deformation regimes for the both portions of the curves in Figure 3 is correct, on the other hand.

Taking into account that the osmotic pressure increases with increasing temperature, we can write equation (6) in the form

$$\Phi_e = P_c / K_{h\ DBP} = C_1 RT / K_{h\ DBP}, \qquad (7)$$

which is qualitatively comparable with 8-12 straight lines in Figure 4.

It should be emphasized that the data presented in Figure 4 are not in agreement with the condition of the thermodynamic equilibrium and, consequently, with equation (7) (the rate of the Φ growth can increase two- or three-fold as a temperature rises by 10°).

However, the decrease in the slopes of the straight lines 8-12 with increasing the DBP content in the films points clearly to the thermodynamic reason for the phenomenon observed. The slope decrease may be caused either by decreasing the concentration C_1 for molecular-claster water dispersed in a sponge in the presence of a great amount of hydrophobic DBP [C_1 appears in the numerator of equation (7)] or by increasing the bulk modulus of elasticity $K_{h\ DBP}$ [it appears in the denominator of equation (7)], if plasticization reduces the compressibility of the sponge under the action of a hydrostatic pressure.

Once the maximum retardation of the structural fluctuations in the plasticized matrix due to compression by growing water drops is attained, the process turns into the regime of viscous flowing the polymer, and does not stop. The point is that as soon as the σ_e value [equation (5)] decreases because of the relaxation displacement of the polymer chains, the fluctuation mechanism, restoring the σ_e value owing to "pumping" water to the growing drops, comes into effect immediately.

Addition of the forces generated by a great number of drops results in specific directionality of the resulting forces - the tangent to the surface ones compress the films, and the normal ones stretch them. Because of this phenomenon, the sample material flows, increasing its thickness and decreasing the area. Such a situation is rather common in actual practice, for example, when breaks and exfoliations appear on coatings of water-emulsion an oil paints.

As was mentioned above, the DBP-free samples of PMMA undergo only the short-time excitement of the structural fluctuations in heated water, which terminate on the completion of the relaxation process of swelling and removing the monomer and MC remains. This fact results from the PMMA transition to the glassy state and from the appearance of structural impediments, which exclude the origination of the expansion and compression waves of the required value, involving the capillary fluctuations of liquid water, inside the polymer sponge. It is just these waves (ensuring the appearance and movement of fluctuation capillary water under conditions of high dynamics) which are responsible for mixing the internal content of polymeric globules and for the separation of impurities consisting of the mobile enough molecules of the MMA and MC type from the polymer.

It should be emphasized that the segregation of drop water in hydrophobic elastomers is always accompanied by the exclusion and redistribution of impurities [3]. In such elastomers, the relationships of this phenomenon closely resembles those observed in plasticized PMMA. From our viewpoint, this resemblance reflects the same in qualitative sense structural-fluctuation properties of the sponge structures constructed of the flexible hydrophobic polymer chains.

REFERENCES

1. Zuev, Yu.S., Polymer Degradation in the aggressive media, Khimia, Moscow, 1972.
2. Rowland, S.P. (ed.), Water in Polymers, D.C., Washington, 1980.
3. Chalykh, A.E., Diffusion in Polymeric Systems, Khimia, Moscow, 1987.
4. Bershtein, V.A., Egorova, L.M., Egorov, V.M. & Sinany, A.B., Visokomol. Soedin., Part A, 31 (1989), 2482.
5. Schegolev, S.Yu. & Klenin, V.I., Opt. i Spektr., 31 (1971) 794.
6. Klenin, V.I. & Schegolev, S.Yu., Visokomol. Soedin., Part A, 13 (1971), 1919.
7. Buehe, F., J. Polym. Sci. (Lett.), 14 (1954) 414.
8. Zhurkov, S.N., Marikhin, V.A. & Slutsker, A.I., Phys. Tverd. Tela, 1 (1959), 1159.
9. Slutsker, A.I. & Marikhin, V.A., Opt. i Spektr., 10 91961) 512.
10. Geguzin, Ya.E., Why and how the empty space is disappearing, Nauka Moscow, 1976.
11. Zaikov, G.E., Iordansky, A.L. & Markin, V.S., Diffusion of Electrolites in Polymers, Khimia, Moscow, 1984.
12. Van Krevelen, D.V., Properties of Polymers. Correlations with Chemical Structure, Elsevier, Amsterdam-London-New York, 1972.
13. Papkov, S.P. & Efimova, S.G., Visokomol. Soedin., 8 (1966) 1875.
14. Volkov, V.V., Bokarev, A.K. & Durgarian S.G., Visokomol. Soedin., Part A, 26 (1984) 1294.
15. Matveyev, Yu.I. & Askadskyi, A.A., Visokomol. Soedin., 28 (1986) 1365.
16. Bezruk, L.I., Lipatov, Yu.S., Ivashenko, V.N., Lipatova, T.E. & Pasechnik, Yu.V., Visokomol. Soedin., Part B, 12 (1970) 35.
17. Topchiev, D.A., Visokomol. Soedin., Part A, 32 (1990) 2243.
18. Mikheyev, Yu.A. & Guseva, L.N., Khim. Phys., 10 (1991) 724.
19. Mikheyev, Yu.A., Intern. J. Polym. Mater., 16 (1992) 221.
20. Hyo-Gun Kim, J. Appl. Polym. Sci., 22 (1978) 889.
21. Kirste, R.G., Kruse, W.A. & Schelten, J., Makromol. Chem., 162 (1972) 299.
22. Ballard, D.G.H., Wignall, G.D. & Scheltn, J., Eur. Polym. J., 9 (1973) 965.

23. Bagley, E.B. & Scigliano, J.M., Polym. Eng. & Sci., 11 (1971) 320.
24. Tager, A.A. & Tsilipotkina, M.V., Uspekhi Khimii, 47 (1978) 152.
25. Butyagin, P.Yu., Uspekhi Khimii, 53 (1984) 1769.
26. Foudas, G., Pakula, T., Stamm, M. & Fischer, E.W., Macromolecules, 26 (1993) 1671.
27. Olkhovik, O.E., Visokomol. Soedin., Part A, 18 (1976) 1012.
28. Chudinov, B.S., Water in wood, Nauka, Novosibirsk, 1984.
29. Bazarov, I.P., Thermodynamics, Vysshaya Shkola, Moscow, 1983, p. 9.
30. Bering, B.P. & Serpinskyi, V.V., Izvest. AN SSSR, ser. khim., 11 (1974) 2427.
31. Bering, B.P., Krasilnikova, O.K. & Serpinskyi, V.V., Dokl. AN SSSR, 231 (1976) 373.
32. Deryagin, B.V., Churaev, N.V. & Muller, V.M., Surface Forces, Nauka, Moscow, 1987.
33. Frolov, Yu.G., Course of Colloid Chemistry, Khimia, Moscow, 1982.
34. Kaye, G.W. & Laby T.H. Tables of Physical and Chemical Constants Longmans, Green & Co., London, New York, Toronto, 1958.

Organosilicon Copolymers with Carbotricyclodecasiloxane Fragments in the Dimethylisiloxane Chain

O.V. Mukbaniani, L.M. Khananashvili, M.G. Karchkhadze, R.Sh. Tkeshelashvili, and N.O. Mukbaniani
I. Javakhishvili Tbilisi State University,
3 Chavchavadze Ave., Tbilisi 380028, Republic of Georgia

The reaction of stage heterofunctional condensation of cis-1,3,5,7-tetrahedroxy-1,3,5,7-tetraphenylcyclotetrasiloxane (tetrol) with 1,3-dichlor-tetramethylisiloxane and 1,3-ichlor-1,3-dimethyl-1,3-divinyldisiloxane by the ration of initial products of 1:1:1 has been studied in the presence of pyridin and 1,3-divinyl-1,3,9,9,111-hexamethyl-5,7,13,15-tetraphenyltricyclodecasiloxane has been obtained.

The reaction of hydrid polyaddition of α,ω-dihydriddimethylsiloxanes to 1,3-divinyl-1,3,9,9,11,11-hexamethyl-5,7,13,15-tetraphyenyl-tricyclodecasil -oxanes in the presence of platinochlorohydric acid and acetylacetonatdicarbonyl rodium catalyst has been studied.

The reaction order, activation energies and hydrid polyaddition rate constants have been found.

It has been established that with an increase of the length (n) of the α,ω-dihydriddimethylsiloxanes the degree of hydrid polyaddition is reduced. It was shown that in the case of acetylacetonatdicarbonyl rodium the obtained copolymers are structured.

Thermogravimetric, thermomechanical and X-ray studies of the synthesized copolymers have been carried out.

INTRODUCTION

It is know that the synthesis of carboorganisoloxane oligomers and polymers is based on the reaction of hydrid polyaddition of organohydridsiloxanes to organoalkenylsiloxanes[1]. It is possible, employing this reaction to obtain the oligomers and polymers of linear structure[2-4], as well as the polymers of cyclolinear structure with eight members fragments in the chain[5,6].

It was shown, that by the introduction of cycocarbosiloxane fragments into dimethylsiloxane chain the thermal-oxidative stability of cyclolinear copolymers increases.

Polycarboxiloxanes with methylcyclotetrasiloxane ethylene-bridged links have been obtained by the reaction of self-polyaddition of 1-hydrid-3-vinylhexamethylcyclotetrasiloxane[7].

The carboxiloxane copolymers comprising in the chain organocyclotetra (penta, hexa) siloxane fragments have been synthesized by the reaction of hydrid polyaddition of dihydroorganocyclotetra (penta, hexa) siloxanes to divinylorganocyclotetra (penta, hexa) siloxanes[8,9]. It was established, that the thermal stability of the copolymers declines with an increase of the volume of the cyclosiloxane ring.

Compared to polyorganosiloxane analogs, the polymers with disilylethylenic groups have lower thermal-oxidative stability, but higher thermal stability in absence of oxygen[8,10].

EXPERIMENTAL PART

The purity of all the initial organochlorosilanes and siloxanes have been controlled by their boiling temperatures, chlor ions content and refraction indexes.

The organic solvents were cleaned by drying and distillation. The initial α,ω-dihydriddimethylsiloxanes have been synthesized by well-known methods [11].

IR spectra of all the samples have been taken on an UR-20 instrument, while the NMR spectra - on a "Perkin Elmer" R-32 at operating frequency of 90 MHz in the solutions of CCl_4 and dioxane HMDS or TMS as internal standard.

The purity of obtained divinyl containing organocyclosiloxane was verified on a gas-liquid chromatograph "ΛXM-8MΔ" phase SKTF-100 (10%), the NAW chromosorb, carrier gas - He, the 2M column.

Diffractograms were taken on a "ΔPOH-2" instrument, A-Cu_α values were measured without a filter, the angular velocity of the motor ω=2 deg/min.

The thermomechanical curves were taken on a custom made installation. The test was carried out as follows: V=5 deg/min, σ=0,1 kg/sm^2.

Thermogravimetric investigations were carried out on a "Paulic-Paulic-Erdey" derivatograph, model MOM-102. The conditions of test were: temperature size rate V=5 deg/min, weight m=100 mg.

Heterofunctional condensation of cix 1,3,5,7-tetrahydroxy-1,3,5,7-tetraphenylcyclotetrasiloxane (tetrol) with 1,3-dichlortetramethyldisiloxane and 1,3-dichlor-1,3-divinyl-1,4-dimethyldisiloxane.

To a solution of 27,6 g (0,05 mole) of tetrol in 500 ml of anhydrous ether at room temperature a solution of 10,1 g (0,05 mole) of the 1,3-dichlortetramethyldisiloxne in 25 ml anhydrous ether and a solution of 7,9 g (0,1 mole) of anhydrous pyridin in 25 ml of anhydrous ether were added.

The reaction mixture was stirred and heated at reflux for 3-4 h. The residue was filtered off, the solvent was evaporated and 36,5 g of a viscous residue was obtained. After distillation of 29,0 g of product in vacuo, 15,7 g (54%) of pure product was isolated, T_{boil}=245-250°C, P=2-3x10-3 mmHg. 14,4 g (92%) 1,3-divinyl-1,3,9,9,11,11-hexamethyl-5,7,13,15-tetraphenyltricyclodecasiloxane was obtained. After recrystallization from the mixture toluene-heptane (30:70). IR bands: 1275 (SiMe); 1000, 1435 (SiPh); 1080 (SiOSi); 2900-3000 (C-H)

Calculated for C34H44Si8O10, %: C 44,76; H 5, 29; Si 26, 83; M=837.

Found: C 48, 62; H 5, 10; Si 27, 07; M=800.

Hydrid polyaddition. The hydrid polyaddition reaction was carried out in three-necked flask equipped with stirrer and reflux condensor with a chlorcalcium tube. The initial reagents were placed into the flask and thermostated in an oil bath until constant temperature was achieved. For the polymer synthesis, 0,01 M solution of platinochlorohydric acid or 0,1 N solution of acetylacetonatdicarbonyl rodium in tetrahydrofuran ($5\div10\text{x}10^{-5}$ g per 1g of the starting substance) was used as a catalyst. The reaction was performed in the argon atmosphere at a 1:1 mole ratio of the starting substances without solvent at temperatures of 70, 80 and 90°C.

The reaction proceeded in dry argon ambient 7-8h. Then the reaction mixture was diluted in toluene, filtered off and precipitated by addition of methyl alcohol.

RESULTS AND DISCUSSION

The present paper deals with synthesis and studies of the properties of carboxiolane copolymers contained in the chain along with dimethylsiloxane fragments tricycloorganodecasiloxane fragments. There are no data in the literature on the carbisiloxane copolymers of the above mentioned structure.

For the purpose of synthesis of 1,3-divinyl-1,3,9,9,11,11-hexamethyl-5,7,13,15-tetraphenyltricyclodecasiloxane (1) the reaction of heterofunctional condensation of cis-1,3,5,7-tetrahedroxy-1,3,5,7-tetraphenylcyclotetrasiloxane with 1,3-dichlortetramethyldisiloxane and 1,3-dichlor-1,3-divinyl-1,3-dimethyldisiloxane was carried out in the presence of pyridin with 1:1:1 ratio of the initial components in the dilutin solution of the anhydrous ether. After distillation and recrystallization of the obtained product 1,3-divinyl-

1,3,9,9,11-11-hexamethyl-5,7,13,15-tetraphenyltricyclodecasiloxane (1) was obtained according to the following scheme:

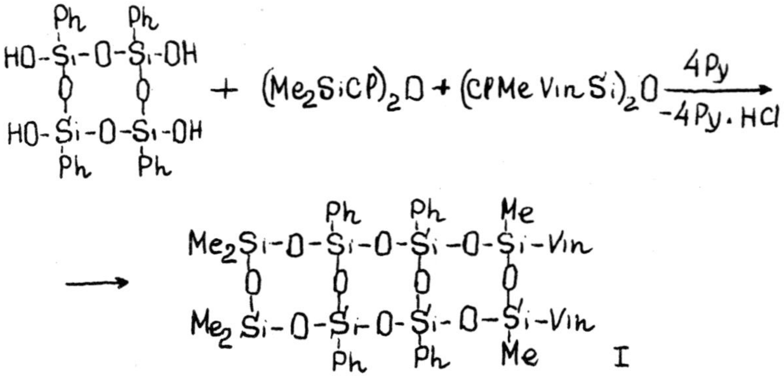

I

The composition and structure of synthesized compound 1 were determined by means of functional and elementary analysis, by finding of molecular mass, by IR and NMR spectra data.

In the ^{1}H NMR spectra of compound 1 one can observe two signals with chemical shifts of δ=0,55 and δ=0,75 ppm, for Si-Me protons, multiplet in the region 7,5÷8,2 ppm for phenyl protons and multiplet for vinyl protons with center of the signal at 6,2 ppm.

In the IR spectra of compound 1 one can observe absorption bands characteristic of asymmetric stretching vibrations of the Si-O-Si bands in the region of 1050 and 1080 cm^{-1}. One can also observe absorption bands for Si-Me, Si-Ph and Si-Vin bonds in the region of 1275, 1435 and 2900-3000 cm^{-1} respectively.

The polymer synthesis was carried out at temperatures not exceeding 90°C, so that cyclosiloxane ring opening would not take place in the conditions of hydrid polyadditions.

Preliminary heating of divinylorganotricyclodescasiloxane (1) within the temperature range of 70-90°C for 10 h in the presence of acetylacetonatdicarbonyl rodium or platinochlorohydric acid as a catalyst showed, that in this conditions polymerization of the starting divinyltricyclodecasiloxane did not occur. The detailed analysis of the reaction mixture by gas-liquid chromatography method has been shown the presence of initial organosiloxanes. Besides, there are no changes in the NMR and IR spectras of compound 1 and dihydriddimethylsiloxanes.

The hydrid polyaddition reaction of compound 1 was carried out employing various length of α,ω-dihydriddimethylsiloxanes. The course of the reaction was watched by decrease of amount of active Si-H groups. It was established, that in the case of using of rodium acetylacetonatdicarbonyl as a catalyst obtained in organic solvents soluble copolymers, which during a time structured. This may be explained that in spite of the polymers reprecipitated from toluene solution by methyl alcohol, catalyst of rodium remained in polymeric systems, which decompose and evoke a decomposition of copolymers. Therefore for receipt of copolymers we used as a catalyst platino chlorohydric acid.

It was observed, that with the increase of length of α,ω-dihydrid-dimethylsiloxane chain the rate and depth of the polyaddition reaction decreases. As it is shown at Figure 1 the conversion of Si-H bond is not completely and decreases from 95% (n=4) to 83% (n=12).

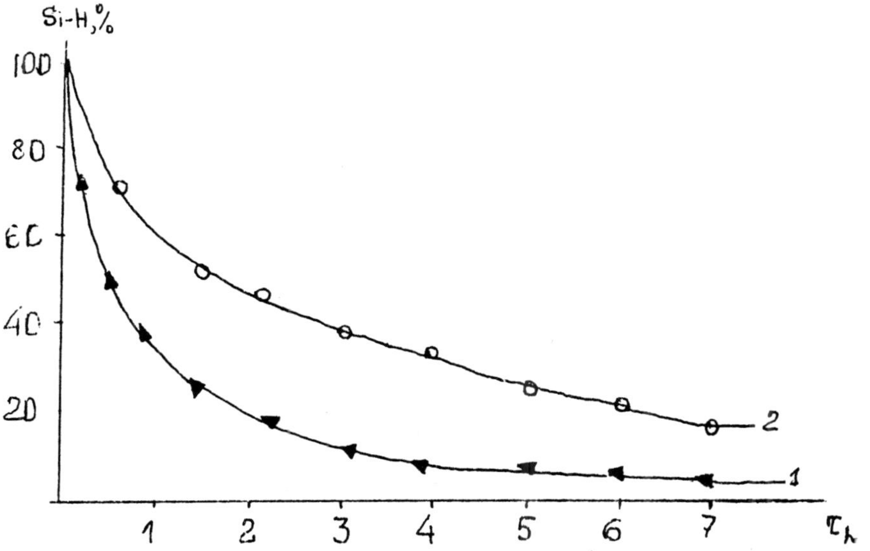

Figure 1. Change in Si-H(%) with time upon polyaddition of α,ω-dihydrid-dimethylsiloxane to divinylorganotricyclodecasiloxane at 90°C, where curve 1 is for n=4, curve 2 is for n=12.

The reaction of hydrid polyaddition proceeds according to the general scheme:

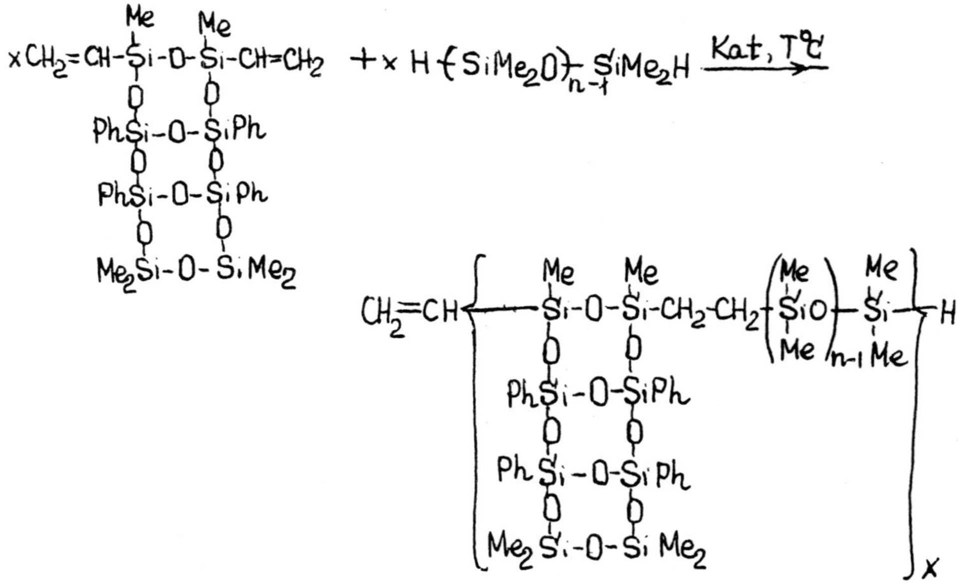

Where n=2 (11), 4(111), 6 (IV), 12 (V), 21 (VI). Catalyst - H_2PtCl_6.

As a result of the reaction, copolymers are obtained with η_{spec}=0,08-0,26, that are liquid or vitrious light yellow transparent products soluble in usual organic solvent. Some physico-chemical properties; elementary composition, molecular weight and yields of synthesized copolymers are presented in the table.

Copolymer	T react, °C	η^*_{spec}	T vitr. °C	d_1, Å	Elementary composition* C	H	Si	Yield %	M*x10^3
II	90	0,08	-12	10,20	47,31 47,01	5,99 5,92	29,01 28,86	80	72
III1	70	0,09	–	–	45,21 45,08	6,31 6,26	29,86 30,05	83	–
III2	80	0,10	–	–	44,96 45,08	6,11 6,26	30,26 30,05	88	–
III3	90	0,11	-50	–	45,31 45,08	6,19 6,26	30,35 30,05	91	85
IV	90	0,14	-96	8,68	43,67 43,60	6,66 6,47	31,10 30,96	92	–
V	90	0,20	-123	–	41,43 41,10	6,86 6,73	32,70 42,51	93	–
VI	90	0,26	-123	7,54	56,61 56,74	4,89 5,08	23,73 23,99	94	110

The course of the reaction was controlled also, by an increase in viscosity of obtained copolymers. It was established that with increase of temperature to 70-90°C the viscosity of copolymers rises.

It was observed, that with temperature rise in the reaction of hydrid polyaddition of compound 1 with α,ω-dihydriddimethylsiloxane the conversion of Si-H bond increases from 85% (70°C) to 95% (90°C).

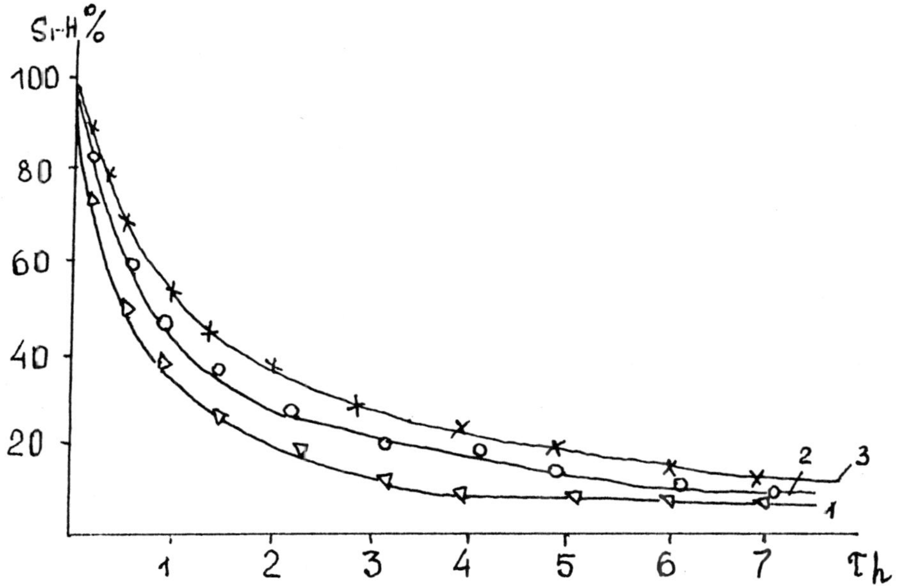

Figure 2. Change in Si-H(%) with time upon polyaddition of α,ω-dihydriddimethylsiloxane (n=4) to divinylorganotricyclodecasiloxane, where curve 1 is for temperature 90°C, curve 2 is for 80°C, curve 3 is for 70°C.

Figure 2 shows dependence of the Si-H% concentration decrease on the time. Figure 3 shows the direct dependence of reverse concentration of the reactant on the time. One can see, that the hydrid polyaddition is the reaction of second order. Proceeding from the data of Figure 3 the reaction rate constants at various temperatures were calculated: $k_{70°C}=1,\ 1086x10^{-2}$, $k_{80°C}=1{,}6196x10^{-2}$, $k_{90°C}=2{,}3834x10^{-2}$. It was found that each time with an increase of the reaction temperature by 10°C the reaction rate constants increases approximately 1,5 times.

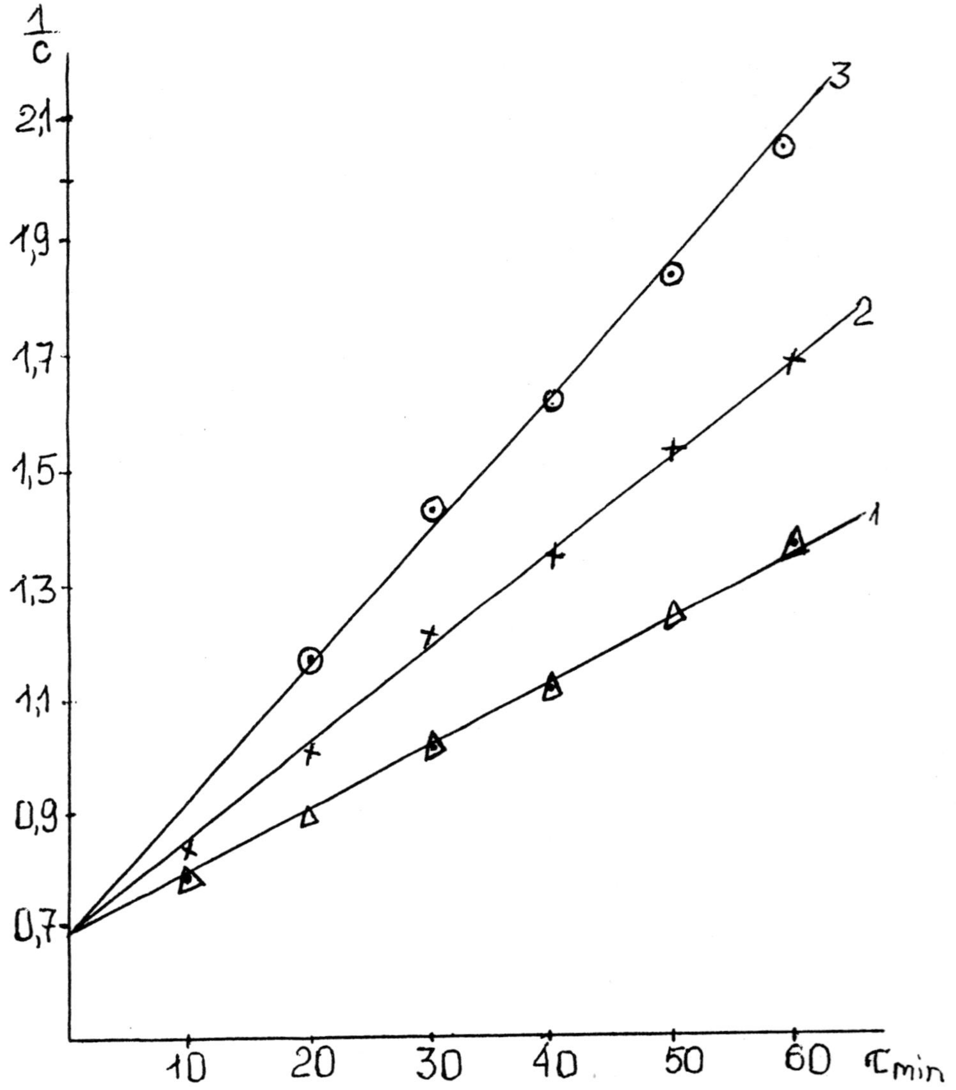

Figure 3. Dependence of reverse concentration on time. Curve 1 is for 70°C curve 2 is for 80°C, curve 3 is for 90°C.

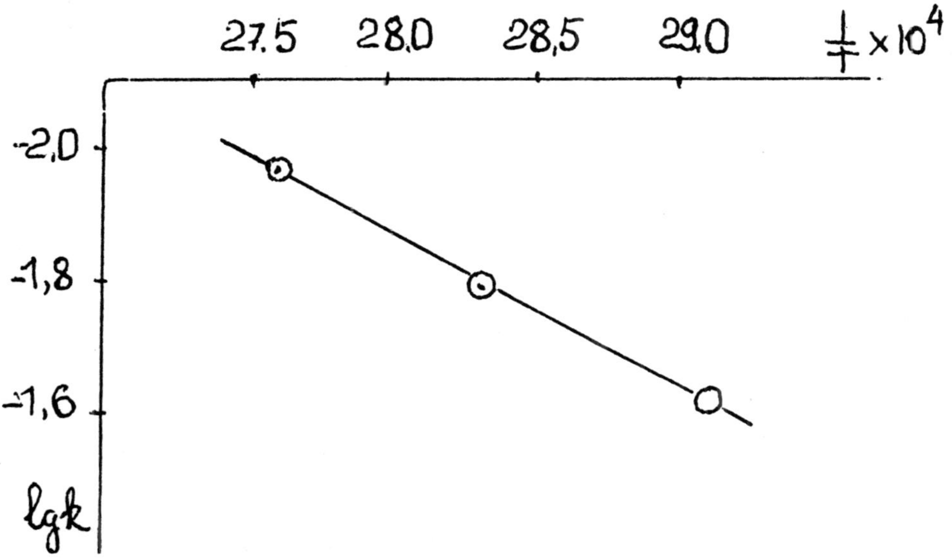

Figure 4. Dependence of reverse temperature on the logarithm of the rate constant of the polyaddition reaction.

Figure 4 shows the dependence of the reaction rate constants logarithm on reverse temperature, from which data the activation energy of the hydrid polyaddition reaction has been calculated to be equal to 64,4 KJ/mole.

The analysis of the IR spectra of the obtained copolymers revealed, that the absorption bands characteristic for Si-H bonds in the range of 2160-2170cm^{-1} do not vanish completely. The new absorption bands appear in the spectra at 1150 cm^{-1}, typical for g fanshaped oscillations of the CH_2 groups, as well as absorption bands in the range of 2900-3000 cm^{-1}, typical for valence oscillations of the CH bonds in the CH_2 groups of the tehylenic bridges.

By ^1H NMR spectra for copolymer 11 it has been established, that catalytic hydrid polyaddition proceeds mainly by the Farmer rule with formation of dimethylenic bridges. In NMR spectrum one can observe a signal for grouping $-CH_2-CH_2-$ with the chemical shift of δ=0,32 ppm. One can observe also a double signal with chemical shift center at δ=1,12 ppm, which can be assigned to methyl proton in the grouping $=CH-CH_3$, whose amount is about 5-6%. Integral relations of methyl and phenyl protons correspond to the formula of the copolymer 11.

Thermomechanical studies of the obtained copolymers have been carried out and it was established that vitrification temperature of copolymers decreases with an increase of the value n of the linear dimethylsiloxane chain (Figure 5). Beginning from value of n=11 the influence of carbotricyclodecasiloxane fragments in copolymers does no more effect on dimethylsiloxane chain, and the vitrification temperature for copolymer V remains of -123°C.

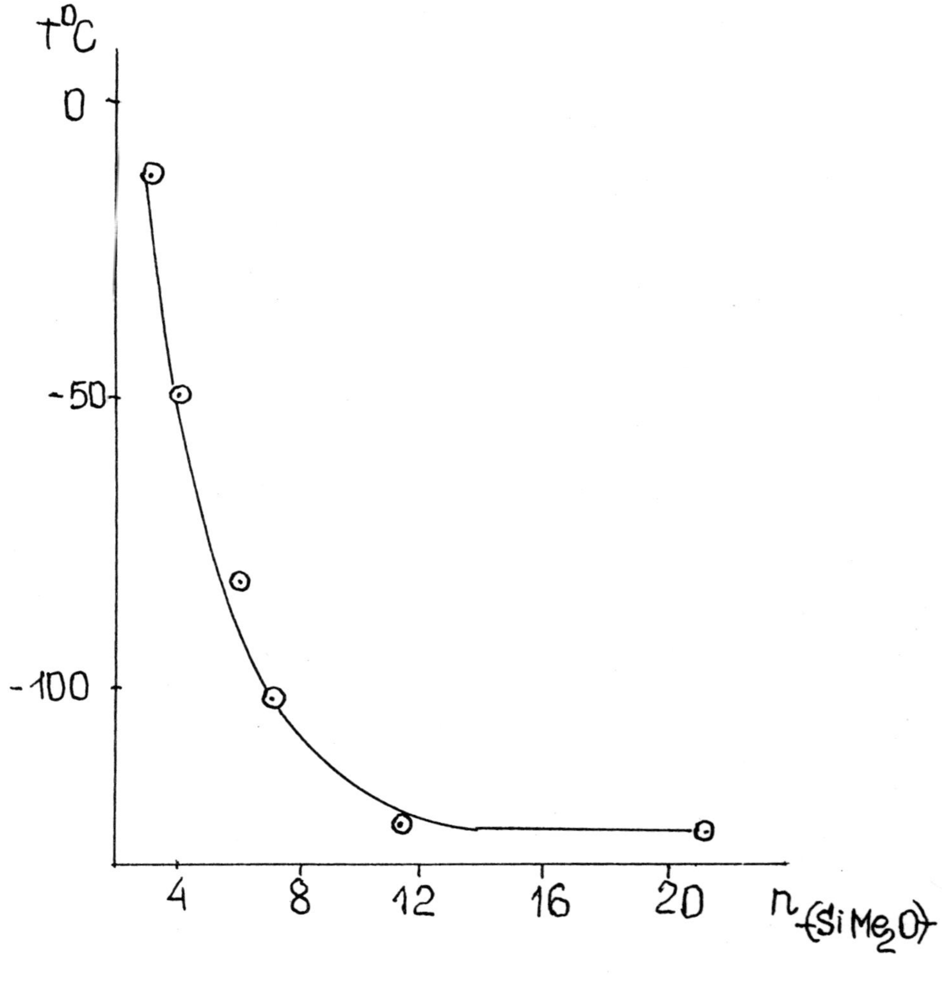

Figure 5. Dependence of vitrification temperature of cyclolinear copolymers on the length n of the linear polydimethylsiloxane.

The increase of the dimethylsiloxane chain length (n=21 does not lead to the formation two-phase system as it was shown early by us[12], in cyclolinear copolymers containing rigid tricyclodecaphenyldecasiloxane and flexible dimethylsiloxane links.

Comparative estimation of vitrification temperature of carbotricyclodecasloxane fragments containing copolymers with tricyclodecasiloxane containing copolymers[12], shows that they are characterized by lower values of vitrification temperature, which may be explained by additional presence of flexible $-CH_2-CH_2-$ groups in copolymers.

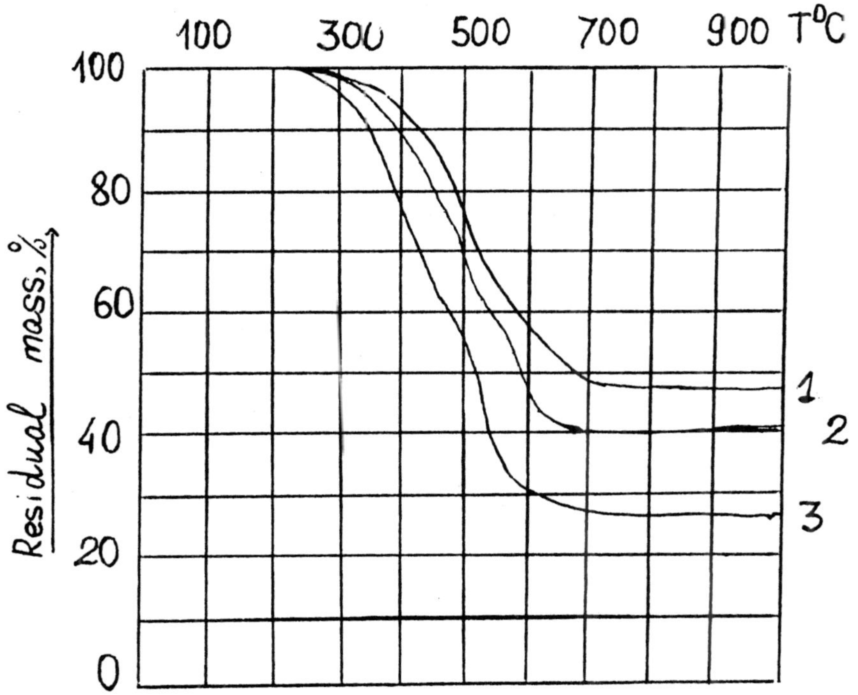

Figure 6. Thermogravimetric curves of cyclolinear carbosiloxane copolymers, where curve I is for copolymer II, curve 2 is for copolymer IV, curve 3 is for copolymer VI (in air, the rate of temperature increase is 5 deg/min).

The thermogravimetric investigation of copolymers has been carried out. These investigations have shown (Figure 6), that in case of low values of dimethylsiloxane chain n, the copolymers are characterized with higher thermooxidative stability. With an increase of the length of dimethylsiloxane chain the thermooxidative stability of copolymers decreases. With their thermooxidative stability carbotricyclodecasiloxane containing copolymers cede to its siloxane analogies[12]. For polymer mass losses in temperature interval 300-350°C do not exceed 3-7% and the main destruction process proceeds in the range 400-650°C above 650°C the curves of mass losses do not change (Figure 6).

By the introduction of carbotricyclodecasiloxane fragments in dimethylsiloxane chain the main destruction process proceeds approximately at 80-100°C higher temperature then the main destruction of unblocked polydimethylsiloxane.

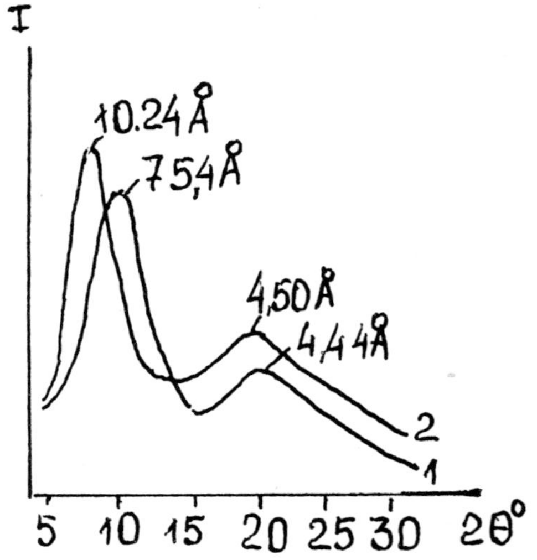

Figure 7. Diffractograms of copolymers, where curve I is for copolymer VI, curve 2 is for copolymer II.

The synthesized copolymers were analyzed by the rentgenographic method. As can be seen from the amorphous polymer diffractograms (Figure 7) in the case of a small values n the interchain distance reached its maximum value d_1=10,24 Å, with an increase of a length dimethylsiloxane chain n=21, interchain distance decreases and for copolymer VI d_1=7,54 Å. So in copolymer VI does not occur formation two-phase systems, as it was observed in copolymers with decaphenyltricyclodecasiloxane fragments in dimethylsiloxane chain, may be explained in our case by combination of the rigid carbotricyclosiloxane and flexible dimethylsiloxane fragments therefore the copolymers are a single-phase systems.

So by means of the reaction of hydrid polyaddition we have synthesized and studied, for the first time, the properties of carbosiloxane cyclolinear copolymers, containing reactionable tricyclosiloxane fragments with regular arrangement in the chain. Such copolymers may be structured without giving of gaseous products.

REFERENCES

1. K.S. Andrianov, I. Souchek and L.M. Khananashvili, Uspekhi Khimii 48, 233 (1979).
2. A.D. Petrov and V.M. Vdovin, Izv. Akad. Nauk SSSR, Otdel Khim. Nauk, 939 (1959); Chem. Abstr., 54, 265b (1960).
3. G. Greber and L. Metzinger, Macromol. Chem., 39, 189 (1960).
4. G. Greber and R. Reese, Macromol. Chem., 77, 13 (1964).
5. K.A. Andrianov, B.S. Tikhonov, I.Yu. Klementiev and M.N. Rozhnova, Vysokomol. Soed., Ser. A, 18, 2288 (1976).
6. A.A. Xhdanov, K.A. Andrianov, and A.P. Malykhin, Dokl. Akad. Nauk, SSSR 211, 1104 (1974).
7. K.A. Andrianov, V.I. Sidorov, M.G. Zaitseva and L.M. Khananashvili, Khim. Heterocycl. Soed., 32 (1967); Chem. Abstr., 67, 73940z (1967).
8. A.A. Zhdanov, and T.V. Astapova, Vysokomol. Soed., Ser. A, 23, 626 (1981), Chem. Abstr., 94, 209261w (1981).
9. O.V. Mukbaniani, L.M. Khananashvili, N.A. Koiava, G.V. Porchkhidze and Yu.I. Tolchinski, Intern. J. Polym. Mater., 17, 113 (1992).
10. V.V. Severni, V.Yu. Flaks, A.A. Zhdanov, V.A. Vlasova, K.A. Andrianov and F.N. Vyshnevski, Vysokomol. Soed., Ser. A, 16, 419 (1974); Chem. Abstr., 81, 28417p (1974).
11. M. Sakiyama and R. Okawara, J. of Organometal. Chem., 2, 473 (1964).
12. K.A. Andrianov, G.L. Slonimski, A.A. Zhdanov, V.Yu. Levin, D.I. Tsvankin, Yu.P. Kvachev, E.S. Obolonkova, N.N. Makarova, O.V. Mukbaniani and E.M. Belavtseva, Vysokomol. Soed., Ser. A, 19, 1507 (1977).

Co-Oligomerization of Thienylcontaining Organic Cyclosiloxanes with 1,3-Bis(3-Glycydoxypropyl) Tetramethyldisiloxanes

L.M. Khananashvili, D.U. Murachashvili, D.A. Girgvliani, Ts.N. Vardosanidze, N.I. Tsomaya, E.G. Markarashvili
Tbilisi State University, 3 Chavchavadze Ave., Tbilisi 380028, Republic of Georgia

Abstract—The co-oligomerization of thienylcontaining organic cyclosiloxanes with 1,3-bis(3-glycydoxypropyl)tetramethyldisiloxanes initiated by the siloxane of tetramethilamonium is studied.

It is found out that the velocity of expenditure of thienylcontaining organocyclosiloxanes is described by the linear equations depending on its instant concentration. The constants of velocity of expenditure of organocyclosiloxanes is calculated and is shown, that they depend on the concentration of 1,3-bis(3-glycydoxypropyl) tetramethyl-disiloxanes in the reaction mixture. The velocity limitation is reached, when th Si-O-Si bound in 1,3-bis(3-gycydoxypropyl) tetramethyldisiloxanes is splitted.

Using the method of gel permeable chromatography it is shown, that in the presence of polar admixture (dimethilphormamide) the system reaches the equilibrium much faster, producing the oligomers of given molecular mass.

Introduction

The investigation of methods of synthesis of oligoorganocyclosiloxanes with carbonfunctional groups with silicon atom is of great interest and is widely investigated by many scientists. This interest is due to the fact, that these compounds can be use to obtain the block co-polymers which can be used in different fields of technology (1-3).

Big reacting ability of siliconorganic compounds with epoxy groups at the ends is the reason of great interest of investigators, and namely, to the oligomers α,ω-bis(3-glycydoxypropyl)diorganodisiloxanes. The main method of obtaining of such oligomers is co-oligomerization of organocyclosiloxanes with 1,3-bis(3-glycydoxypropyl)tetramethyldisiloxanes (I) in the presence of nucleophilic initiator - silanolate of tetramethylamonium (II) (4).

The kinetics of anionic co-polymerization of cyclosiloxanes containing dimethyl, methylphenyl and triftorpropyl groups to the silicon atoms are described in refs. (5-7). The investigations searching for the possibility to improve the properties of polyorganic siloxanes by the change of part of alkyl- and arylsiloxane pieces by other organic siloxane pieces are also of great interest. In particular the synthesis of organoxiloxane oligomers, containing thienyl group bounded to silicon atom, together with organic radicals, is very interesting (8).

In the present paper we have investigated the influence of concentration of initiator (II) and dimethylformamid (DMFA) used as aproton addition, on the content of products and the speed of reaction of co-oligomerization (I) with 1,3,5-trimethyl-1,3-divinyl-5-thienyl(D^2D^{Th}) cyclotrisiloxanes, and also 1,3,5,7-tetramethyl-1,3,5-trivinyl-7-thienyl-($D_3^{vin}D^{Th}$), 1,1,3,3,5,7-hexamethyl-5,7-dithienyl($D_2D_2^{Th}$) and 1,3,5,7-tetramethyl-1,3,5,7-tetrathienyl D_4^{Th})cyclotetrasiloxanes.

The organosiloxanes of following structure were synthesized:

$$CH_2CHCH_2O(CH_2)_3\overset{CH_3}{\underset{CH_3}{Si}}-\left[O\overset{R}{\underset{R'}{Si}}\right]_n-\left[O\overset{R''}{\underset{R'''}{Si}}\right]_m-O\overset{CH_3}{\underset{CH_3}{Si}}-(CH_2)_3OCH_2CHCH_2$$

$$n=4, m=2, R=R''=Me, R'=Vin, R'''=Th\ (III, III')$$

$$n=8, m=2, R=R''=Me, R'=Vin, R'''=Th(IV, IV')$$

$$n=4, m=3, R=R''=Me, R'=Vin, R'''=Th(V, V')$$

$$n=8, m=3, R=R''=Me, R'=Vin, R'''=Th(VI, VI')$$

$$n=8, m=1, R=R'=R''=Me, R'''=Th(VII, VII')$$

$$n=16, m=1, R=R'=R''=Me, R'''=Th(VIII, VIII')$$

$$n=16, m=0, R=Me, R'=Th(IX, IX')$$

$$n=32, m=0, R=Me, R'=Th(X, X')$$

The oligomers III'-X' were obtained in the presence of 1 mole % DMFA.

EXPERIMENTAL

Co-polymerization was held at the temperature 80°C, in argon surround, by the thermostation of reaction mixture, containing (I) and $D_2^{vin}D^{Th}$,

$D_3{}^{vin}D_{Th}$, D_2D^{Th} and $D_4{}^{Th}$ correspondingly, and also initiator (II), by stirring till the equilibrium was achieved.

After the reaction was finished, the obtained products were filtered and warmed till 140-150°C, during 2-3 hours. To extract the low molecular cyclosiloxanes, the reaction mixture was cooled, and then vacuumed at the temperature 130°C and pressure 266 Pa.

The reaction process was controlled by the change of content of thienil-containing organosiloxanes in the reaction mixture by the method of GC using the chromatograph LXM-80 (model 2). The tube was 3000x4mm, nozzle-chromosorbe W, phase 5mass%, SE-30 (gas carrier was helium), and by change of 1% solution of reaction mixture in toluol at temperature 20°C, which was measured by the viscosimeter of Ubbelode.

The oligomeric products were investigated by the method of gas vapor chromatography (GVC) on liquid chromatograph "Knauer" (the lichrosorb tube was Si-60 and Si-300, the solvent was methilethileton (MEK) (1ml/min), and detector was the differential refractomer c=0.02g/min).

RESULTS AND DISCUSSIONS

When the concentration of initiator (II) is less than $5x10^{-2}$ mole/l, the co-oligomerization of compound (I) with thienylorganocyclosiloxanes proceeds slowly and the final conversion with respect to cyclosiloxanes is lower than of the equilibrium one (P_{eq}=95-96%). That's why the further investigation of reaction was held for the concentration of initiator (II) equal to $5x10^{-2}$ mole/l.

In the process of co-oligomerization of thienylorganocyclosiloxanes with compound (I), incited by (II), the conversion with respect to organocyclosiloxanes reaches 900-94% after 30-60 min (Figure 1a). At the same time the maximum value of η_{sp} is achieved (Figure 3a) which is preserved during the 10-12 hours, and then decreases till the constant value. The velocity of expenditure of $D_2{}^{vin}D^{Th}$, $D_3{}^{vin}D^{Th}$, D_2D^{Th} and $D_4{}^{Th}$ is described by the linear equations depending on its current concentration (Figure 1b, table).

The constants of velocity of co-oligomerization of thienylcontaining organosiloxanes with compound (I), increases according to the following scheme: $D_3{}^{vin}D^{Th}$<$D_2{}^{vin}D^{Th}$<$D_2D_2{}^{Th}$. The higher values of velocities of co-oligomerization of $D_4{}^{Th}$ and $D_2D_2{}^{Th}$ in comparison with $D_2D_4{}^{Th}$ and $D_3{}^{vin}D^{Th}$ can be explained by the difference in number of thienyl-groups in these compounds. During the co-polymerization of compound (I) with $D_2{}^{vin}D^{Th}$ and $D_3{}^{vin}D^{Th}$, incited by (II), the ionic pairs are produced, when silicon bounds of cyclosyloxanes are splitted by the initiator.

Table. The conditions of synthesis, physical and chemical properties of oligosiloxanes and kinetic constants of the reaction of co-oligomerization.

No	DMFA mass %	TIME hours	η_{sp}	M=[η* 5*10^-3	M (ebu-lioscopy)	The concentration of epoxi-groups mass %	Constants of velocity Kmin^{-1}	a	Correlation coefficient r^2
III	0	35	0.031	2081	2200	4.01	0.051	0.037	0.997
III'	1	13	0.029	1881	1900	4.52	0.278	-0.140	0.987
IV	0	35	0.048	4036	4200	2.15	0.049	0.035	0.998
IV'	1	12	0.042	3297	3400	2.52	0.243	-0.120	0.990
V	0	44	0.033	2288	2400	3.66	0.061	0.041	0.999
V'	1	13	0.031	2081	2200	3.91	0.350	-0.065	0.987
VI	0	45	0.050	4294	4400	2.00	0.058	0.053	0.999
VI'	1	12	0.047	3909	4000	2.15	0.307	-0.061	0.995
VII	0	30	0.035	2502	2500	3.44	0.084	-0.0002	0.997
VII'	1	12	0.330	2288	2300	3.74	0.430	-0.048	0.994
VIII	0	35	0.050	4294	4100	2.05	0.073	0.014	0.996
VIII'	1	11	0.048	4037	3950	2.18	0.380	-0.087	0.997
IX	0	32	0.041	3179	3200	2.70	0.110	-0.0005	0.982
IX'	1	10	0.038	2833	2900	2.97	0.720	-0.049	0.965
X	0	32	0.060	5660	5750	1.51	0.091	0.048	0.995
X'	1	11	0.057	5237	5400	1.59	0.667	-0.091	0.968

The epoxigroups of compound (I) can act as aproton ester groups, resulting the improvement of activity of initiator (II). The comparison of constants of velocities for oligomers III-X and III'-X' shows, that they depend on partial content of initial reagents, and so, on concentration of compound (I). This fact proves, that the epoxigroups improve the activity of initiator (II).

The activity of initiator is improved even more in the presence of DMFA. That's why the co-oligomerization of thienilorganocyclosiloxanes with compound (I), incited by the compound (II), in the presence of 1mass% of DMFA, proceeds much more quicker.

During the co-oligomerization of thienylorganocyclosiloxanes with the compound (I), in the presence of DMFA, the conversion of cyclosiloxanes reaches 93-94% already after 5-15 minutes from the beginning of reaction (Figure 2a). The velocity of expenditure of organocyclosiloxanes are also described by the linear equations depending on the current concentration (Figure 2b). The constants of velocities are given in table.

The change of η_{sp} of reacting mixture in the presence of DMFA has also an extremal character (see Figure 3b).

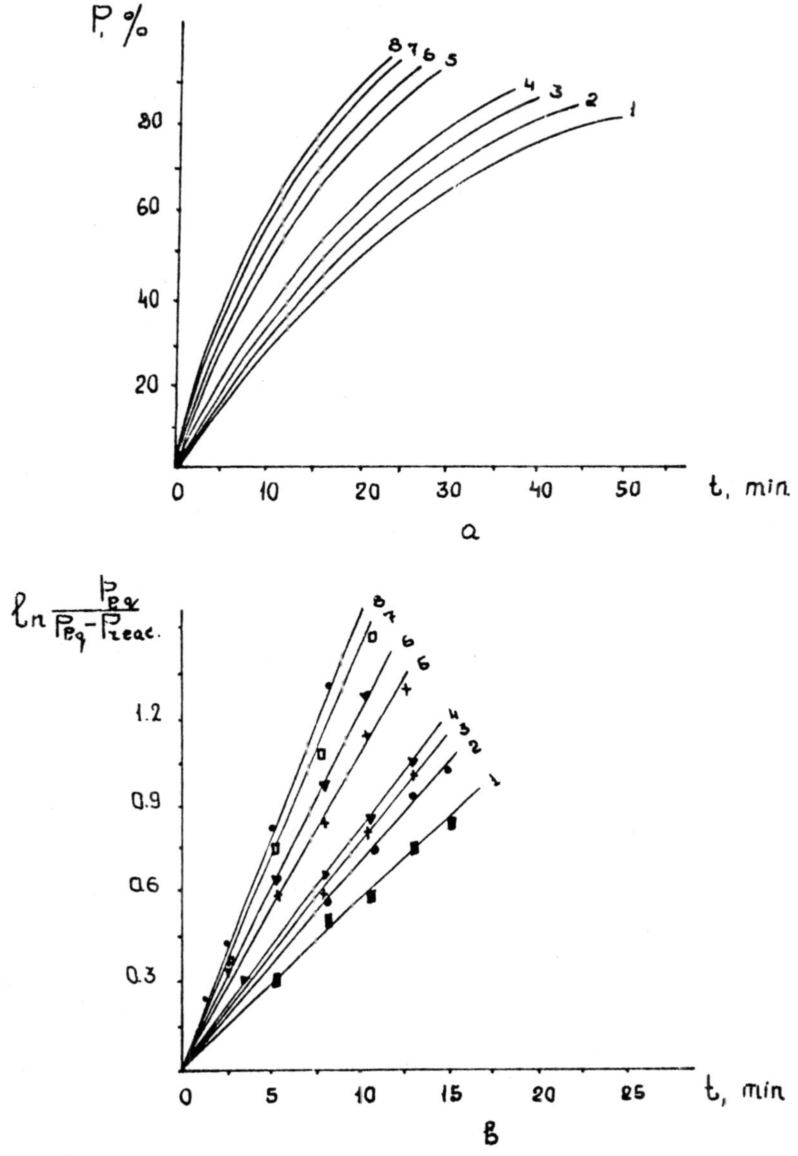

Figure 1. The dependence of conversion (a) and $\ln \frac{P_{eq}}{P_{eq} - P_{reac}}$ (b) of siloxanes from the time of co-oligomerization of compound (I) with $D_3^{vin}D^{Th}$(1,2), $D_2^{vin}D^{Th}$(3,4), D_2D^{Th}(5,6), D_4^{Th} (7,8) in the rate 1:4 and 1:8 indicated by the compound (II) without an addition of DMFA.

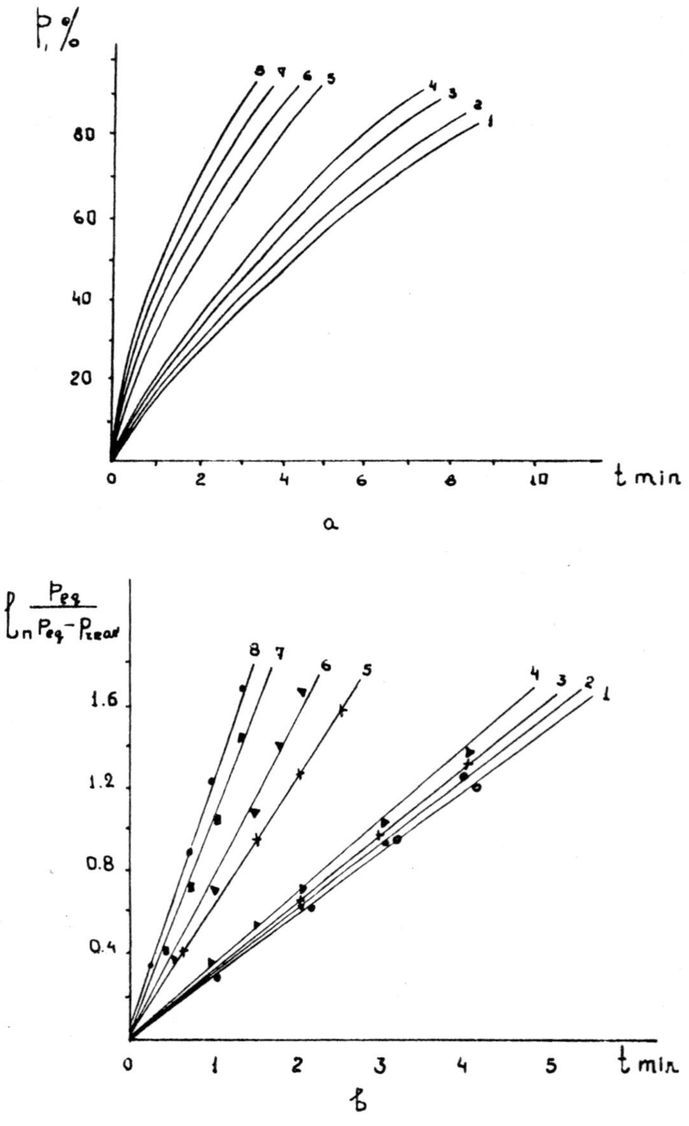

Figure 2. The dependence of conversion (a) and $\ln \frac{P_{eq}}{P_{eq} - P_{reac}}$ (b) of siloxanes from the time of co-oligomerization of compound (I) with $D_3^{vin}D^{Th}$(1,2), $D_2^{vin}D^{Th}$(3,4), D_2D^{Th}(5,6), D_4^{Th}(7,8) in the rate 1:4 and 1:8 initiated by the compound (II) with an addition of DMFA.

Directly after the addition of compound (II) the growth of η_{sp} begins, which achieves its maximal value and then decreases till the equilibrium state.

As can be seen from Figure 3b, when the constant expenditure of organosiloxanes is achieved, the molecular masses of produced oligomers are higher, than an equilibrium one. This fact shows, that the reacting ability of compound (I) is lower than of thienilorganocyclosiloxanes.

The decrease of η_{sp} after the reaching of maximum, shows that in this case the decrease of molecular masses of reacting products takes place. As the content of cyclic products is not changed, the observed decrease of molecular masses of reacting products is due to the fact of the decrease of average weight of molecular masses, with the preservation of average number of the later, due to the interaction of initial compound (I) with the produced silicon chains.

The investigation of products of co-oligomerization of D_4^{Th} and $D_2D_2^{Th}$ with compound (I), by the method of gas vapor chromatography shows, that during the whole process the bimodal distribution of reacting mixture is valid. One maximum corresponds to the low molecular organosiloxanes and initial compound (I). During the reaction its value decreases. Second maximum corresponds to the oligomeric products, and increases during the reaction (Figure 4).

For the oligomers (III') and (X') the mass values M_n, M_w and M_r for oligomeric parts achieve the following values: 5425, 11393, 21077 and 4961, 9823, 17681, corresponding to the maximal value of η_{sp}, and then decreases till minimal value (4200, 7980, 14364 and 4100, 8155, 14516 correspondingly) with the constant value of η_{sp}.

The ratio of M_w/M_n for oligomeric components of reaction products as during the reaction process as after its end is close to two. This value is characteristic to the equilibrium reaction of oligomerization of organocyclosyloxanes with big contribution of interchain interactions (9).

The analysis of gas vapor chromatography shows that without an addition of DMFA, the total content of low molecular products, after the achievement of constant value of η_{sp} is equal to 30-35 mass%.

After the addition of DMFA the content of low molecular products is only 15-17mass%, which is almost the value of the state of equilibrium. This fact indicates that the reaction mixture achieves the equilibrium state when the polar admixture is used.

The values of molecular masses of oligomers obtained as well as in the presence of DMFA, so without it, after the extraction of low molecular components, was determined by the ebulioscopic method, by the presence of cyclic epoxi groups [10] (see table).

The reaction of co-oligomerization with the consideration of all stages of process can be described as:

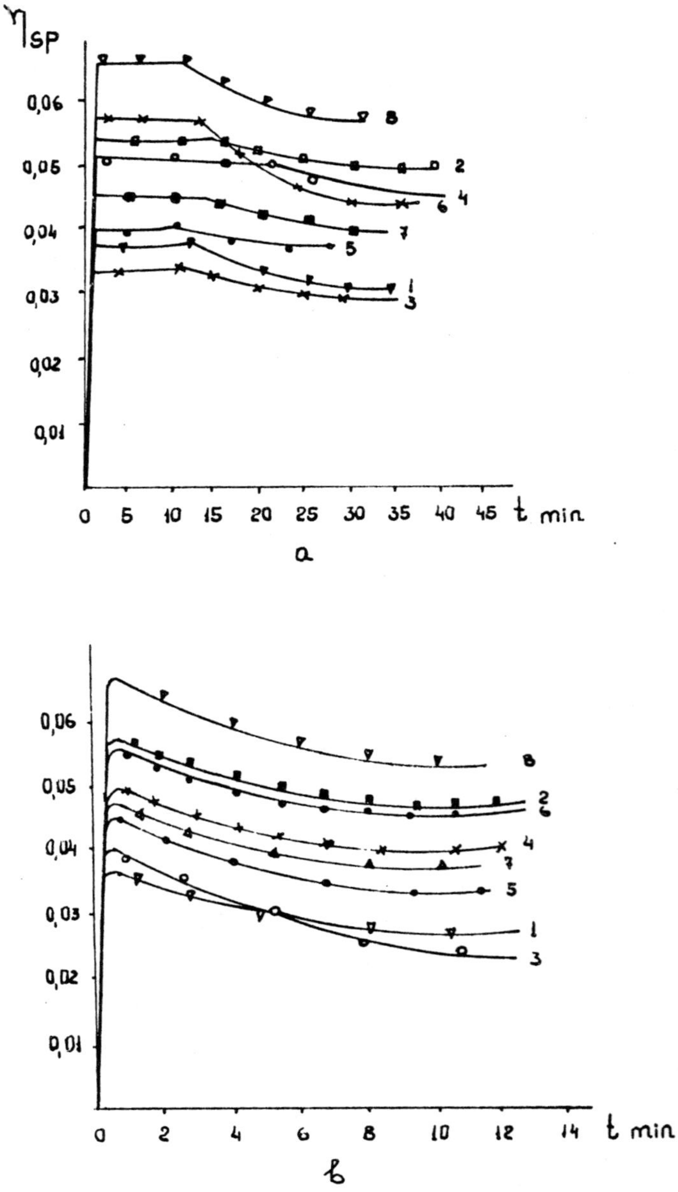

Figure 3. The dependence of η_{sp} from the time of co-oligomerization of compound (I) with $D_3^{vin}D^{Th}$(1,2), $D_2^{vin}D^{Th}$(3,4), D_2D^{Th}(5,6), D_4^{Th} (7,8) in the rate of 1:4 and 1:8 without (a) and with (b) an addition of DMFA.

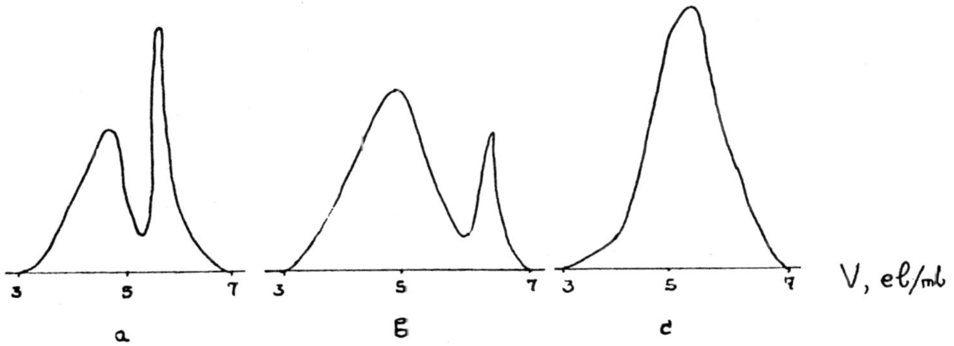

Figure 4. The change of molecular mass distribution of co-oligomerization of products D_2D^{Th} (rate 1:8) initiated by the compound (II) with an addition of DMFA. The reaction time is 1min (a), 5 min (b), 8 hours (c).

$$\begin{aligned}&\sim Si(CH_3)_2OW(CH_3)_4 + XSi(CH_3)_2OSi(CH_3)_2X \Leftrightarrow \\ &\Leftrightarrow \sim Si(CH_3)_2OSi(CH_3)_2 - X + XSi(CH_3)_2ON(CH_3)_4\end{aligned} \qquad 1.$$

$$\begin{aligned}&\Leftrightarrow xSi(CH_3)_2ON(CH_3)_4 + P[RR'SiO]_n \Leftrightarrow \\ &\Leftrightarrow xSi(CH_3)_2O[SiRR'O]_m N(CH_3)_4 + a[RR'SiO]_b \\ &xSi(CH_3)_2O[SiRR'O]_{m+1}Si(CH_3)_2 - x + xSi(CH_3)_2ON(CH_3)_4 \Leftrightarrow \\ &\Leftrightarrow xSi(CH_3)_2O[SiRR'O]_{m-q} - Si(CH_3)_2 - x + \\ &+xSi(CH_3)_2[SiRR'O]_{q+1} - N(CH_3)_4\end{aligned} \qquad 2.$$

where X=CH_2-CH $CH_2O(CH_2)_3$ - R and R' are given below.
O

In the infrared spectrum of synthesized compounds III-X and III'-X' the characteristic adsorption stripes of epoxy groups in the range of 3060 and 915 cm^{-1} are present. The strong stripes at 705 and 1220 cm^{-1}, medieval stripes at 500, 1520, and 1800 cm^{-1}, week stripe at 3035 cm^{-1}, characterizes the presence of C-S bounds (thienil groups).

Strong adsorption stripes at 1080-1060 cm^{-1}, indicates the presence of SiOSi bounds in compounds III-X and III'-X'

In the PMR spectra of compounds III-X and III'-X' the resonance signals can be detected at 7.06-7.78 m.d. (the thienil group).

As for the proton signals of epoxy cycle, they show as complex multiples in the area 2.4-2.9 m.d. The signals in th area 5.2-6.1 m.d. belongs to the protons of vinyl groups.

REFERENCES

1. Yilgor I. Riffle J.S. McGrath J.E.//Ann. Chem. Soc. Symp. Ser. 1985. No 282. pp. 161-174.
2. Coquert X. Lablache-Combier A. Loucheux G. // Eur. Polym. J. 1988. Vol. 24. N. 12. pp. 1137-1143.
3. Gagnebien D. Madec P.J. Marechal E. // Eur. Polym. J. 1985. Vol. 21. N. 3. pp. 273-287.
4. Shapatin A.S. 1127887 USSR. RJCh. 1985 10H 109P.
5. Riffle J.S. Yilgor I. Tran C. Wilkes G.L. McGrath J.E. // Am. Chem. Soc. Symp. Ser. 1983. N 221. pp. 21-53.
6. Yorkgitis E.M. eiss N.S. Tran C. Wilkes G.L. McGrath J.E. // "Epoxi Resins and Composites. I" Berlin. 1985. pp. 79-109.
7. Murachashvili D.U., Kopilov V.M., Khananashvili L.M., Shkolnik M.I. Tsomaya N.I., Volkova R.V., Savitzkiy A.A. // High mol. Comp. 1989. v 31. N 3. Pp 168-172.
8. Khananashvili L.M., Chernisheva E.A., Vardosanidze Ts.N., Markarashvili E.G., Kantaria M.G., Savushkina V.I., Labartkava M.O. // J. General Chemistry. 1984. v. 53. Pp. 1575-1578.
9. Rafikov S.R. Budtov V.P. Monakov U.B. // Introduction in the Physical Chemistry of polymere solutions (in Russian) M. 1978. p. 205.
10. Jay R.R. // Analyt. Chem. 1964. Vol. 36. p. 667.

Subject Index